Nelson Advanced Science

Transition Metals, Quantitative Kinetics and Applied Organic Chemistry

Brian Chapman

First published in 2001 by:
Nelson Thornes (Publishers) Ltd

Nelson Thornes Ltd
Delta Place
27 Bath Road
CHELTENHAM
GL53 7TH
United Kingdom

02 03 04 05 / 10 9 8 7 6 5 4 3

A catalogue record for this book is available from the British Library

ISBN 0 17 448292-2

Illustrations and typesetting by Hardlines, Charlbury Oxford
Picture Research by Zooid Pictures Limited

Printed and bound in China by L.Rex Printing Co., Ltd.

Acknowledgements
The authors and publisher are grateful for permission to include the following copyright material:

Photographs
Alan Thomas: figures 3.2, 3.3, 5.6, 5.9
British Steel PLC: figure 1.11
Chris Fairclough: figure 8.1bl
Corbis UK Ltd: David Katzenstein/Corbis UK lTd: figure 8.1tl, Roger Ressmeyer/Corbis UK Ltd; 8.13
High-Light Protographic, Eaglescliffe/Cleveland/Shell: figure 8.10
Holt Studios International: figures 8.6, 8.7, 8.9
Martyn F. Chillmaid: figure 8.12
Peter Gould: figures 7.6, 7.7, 7.8, 7.10, 7.11, 7.12
Planet Earth Pictures: John Downer/Planet Earth Pictures; figures 8.8
Science Photo Library: figures 3.1, Geoff Tomkinson/SPL; 8.1r, front cover, back cover, John Mead/SPL; 1.9

Every effort has been made to trace all the copyright holders, but where this has not been possible the
publisher will be pleased to make any necessary arrangements at the first opportunity.

Contents

Introduction

Introduction

This series has been written by Chief Examiners and others involved directly with the development of the Edexcel Advanced Subsidiary (AS) and Advanced (A) GCE Chemistry specifications.

Transition Metals, Quantitative Kinetics and Applied Organic Chemistry is one of four books in the Nelson Advanced Science (NAS) series developed by updating and reorganising the material from the Nelson Advanced Modular Science (AMS) books to align with the requirements of the Edexcel specifications from September 2000. The books will also be useful for other AS and Advanced courses.

Transition Metals, Quantitative Kinetics and Applied Organic Chemistry provides coverage of Unit 5 of the Edexcel specification. A study of reduction and oxidation in terms of reduction potentials prepares the way for the extension of inorganic chemistry into the transition metals. Quantitative kinetics does not make difficult mathematical demands on the student; determination of order is essentially based on the comparative study of initial rates. The first chapter on organic chemistry begins by establishing the special nature and properties of benzene as an aromatic compound. A synoptic study of mechanism is then followed by a more relaxed look at some commercial aspects of chemistry. A special feature of the new edition is the full recognition of the importance of spectroscopy to structure determination. This is illustrated by many examples of the use of MS, IR, NMR and UV in the solution of problems and the provision of such problems for the student to solve.

Other resources in the series

NAS Teachers' Guide for AS and A Chemistry provides advice on safety and risk assessment, suggestions for practical work, key skills opportunities and answers to all the practice and assessment questions provided in *Structure, Bonding and Main Group Chemistry; Organic Chemistry, Energetics, Kinetics and Equilibrium; Periodicity, Quantitative Equilibria and Functional Group Chemistry; and Transition Metals, Quantitative Kinetics and Applied Organic Chemistry.*

NAS *Make the Grade in AS and A Chemistry* is a Revision Guide for students. It has been written to be used in conjunction with the other books in this series. It helps students to develop strategies for learning and revision, to check their knowledge and understanding and to practise the skills required for tackling assessment questions.

Features used in this book

The Nelson Advanced Science series contains particular features to help you understand and learn the information provided in the books, and to help you to apply the information to your coursework.

The following are the features that you will find in the Nelson Advanced Science Chemistry series:

INTRODUCTION

Text encapsulates the necessary study for the Unit. Important terms are indicated in **bold**.

5 Oxidation/reduction: an introduction

Introduction

Oxidation and reduction are found with all but four elements in the Periodic Table, not just with the transition metals, although they show these reactions to such an extent that they could be accused of self-indulgence.

When magnesium reacts with oxygen (Figure 5.1)

$$2Mg(s) + O_2(g) \rightarrow 2MgO(s)$$

the product contains Mg^{2+} and O^{2-} ions. Reaction with oxygen is pretty clearly oxidation. The reaction of magnesium with chlorine

$$Mg(s) + Cl_2(g) \rightarrow MgCl_2(s)$$

gives a compound with Mg^{2+} and Cl^- ions. In both cases the magnesium atom has lost electrons, so as far as the magnesium is concerned the reactions are the same. This idea is generalised into the definition of oxidation as loss of electrons. Reduction is therefore the gain of electrons. Since electrons don't vanish from the universe, oxidation and reduction occur together in **redox** reactions.

Oxidation numbers

For simple monatomic ions such as Fe^{2+} it's easy to see when they are oxidised (to Fe^{3+}) or reduced (to Fe). For ions such as NO_3^- or SO_3^{2-} which also undergo oxidation and reduction it is not always so easy to see what is happening in terms of electrons. To assist this, the idea of **oxidation number** or **oxidation state** is used. The two terms are usually used interchangeably, so that an atom may have a particular oxidation number or be in a particular oxidation state.

DEFINITION

Oxidation is electron loss.
Reduction is electron gain.

MNEMONIC

OIL RIG:

oxidation **i**s **l**oss

reduction **i**s **g**ain

Figure 5.1 The use of magnesium flares in photography being demonstrated at an early meeting of the British Association in Birmingham (1865).

Definition boxes in the margin highlight some important terms.

DEFINITION

First ionisation energy: the amount of energy required per mole to remove an electron from each atom in the gas phase to form a singly positive ion, that is

$$M(g) \rightarrow M^+(g) + e^-$$

Second ionisation energy: the energy per mole for the process

$$M^+(g) \rightarrow M^{2+}(g) + e^-$$

and so on for successive ionisation energies.

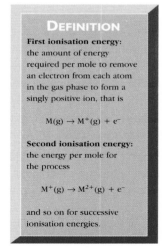

Questions in the margin will give you the opportunity to apply the information presented in the adjacent text.

The **empirical formula** shows the ratio of atoms present in their lowest terms, i.e. smallest numbers. Any compound having one hydrogen atom for every carbon atom will have the empirical formula CH; calculation of the **molecular formula** will need extra information, since ethyne, C_2H_2, cyclobutadiene, C_4H_4, and benzene, C_6H_6, all have CH as their empirical formula. Empirical formulae are initially found by analysing a substance for each element as a percentage by mass.

> **QUESTION**
>
> Find the empirical formula of the compound containing C 22.02%, H 4.59%, Br 73.39% by mass.

Practice questions are provided at the end of each chapter. These will give you the opportunity to check your knowledge and understanding of topics from within the chapter.

Assessment questions are found at the end of the book. These are similar in style to the assessment questions for Advanced GCE that you will encounter in your Unit Tests (exams) and they will help you to develop the skills required for these types of questions.

Questions

1   Plot on a (small) graph the first ionisation energies of the elements from sodium to argon, and account for the shape obtained.

2   Use data from a data book to plot a graph of atomic radius vs atomic number for the elements of Periods 2 and 3 (Li to Ar). Account for the difference in the graphs between Groups 2 and 3.

3   Sketch the structures of:
    (a) the giant covalent lattice of silicon
    (b) the molecule P_4
    (c) the molecule S_8.

4   Silicon has no compounds in which the silicon atom forms double bonds with other elements. Phosphorus, by contrast, does form double bonds with other elements. Suggest why silicon and phosphorus are different in this respect.

About the Author

Brian Chapman was formerly Head of Science at Hardenhuish Comprehensive School until he retired. He is a Chief Examiner in Chemistry for Edexcel.

Acknowledgements

The authors and publisher would like to thank Geoff Barraclough for his work as Series Editor for the original series of four NAMS books, from which the new suite of NAS books was developed. Thanks also to Ray Vincent, Assessment Leader at Edexcel, who checked both the text and the assessment questions for the NAS student books.

I am grateful to NIST for IR, MS and UV spectra and to SDBS for IR, MS and NMR spectra, which I downloaded from their web sites.

Redox equilibria

More about oxidation numbers

You have already met some aspects of oxidation – from the unsophisticated and limited idea of combination of a species with oxygen, to the generally applicable concept of removal of electrons from a species. Since electrons must 'go somewhere' in oxidation, oxidation must always be accompanied by the opposite process – reduction – in which electrons are added.

Where charges and electron transfers are not obvious, we can express the extent of oxidation and reduction by the use of oxidation numbers and their charges. Let's begin this chapter by looking at oxidation numbers, and then comparing the oxidising powers of different oxidising agents for reactions in aqueous solution.

For simple monatomic ions such as Fe^{2+} it's easy to see when they are oxidised (to Fe^{3+}) or reduced (to Fe). For ions such as NO_2^- or SO_3^{2-}, which also undergo both oxidation and reduction, it is not always so easy to see what is happening in terms of electrons. To assist this, the idea of **oxidation number** (or oxidation state) is used. The two terms are usually used interchangeably, so that an ion may have a particular oxidation number or may be in a particular oxidation state.

Each element in a compound is treated as though it is an ion, no matter what the actual nature of the bonding. During a reaction if the 'charge' on the 'ion' becomes more positive, then that part of the compound has been oxidised. The inverted commas are used because the compound may not be ionic – it is taken to be so for this electronic book-keeping exercise.

Some atoms have defined oxidation states. There are three rules:

1. Uncombined elements have an oxidation number of zero.

2. A simple monatomic ion has an oxidation number that is the same as its charge. The oxidation number is given using common numerals (Arabic: 1, 2, 3…) with the appropriate sign, except when naming compounds, where Roman numerals (I, II, III, IV…) are used. Thus Fe^{3+} is iron(+3), but a compound of it would be, say, iron(III) chloride.

3. In compounds, the oxidation number of hydrogen is (+1), and that of oxygen is (–2) (there are some exceptions, considered later); fluorine is **always** (–1).

QUESTION

What is the oxidation number of nitrogen in NO_2, NH_3, N_2, NH_4^+(aq), NO_3^-?

Consider the reaction between bromide ions and chlorine. The oxidation numbers are shown underneath each substance:

$$Cl_2 \text{ (aq)} \quad + \quad 2Br^-\text{(aq)} \quad \rightarrow \quad 2Cl^-\text{(aq)} \quad + \quad Br_2\text{(aq)}$$
$$\text{(0)} \qquad\qquad \text{(–1)} \qquad\qquad\quad \text{(–1)} \qquad\qquad \text{(0)}$$

REDOX EQUILIBRIA

QUESTION

A sample of cast iron of mass 0.50 g was converted to an acidified solution of iron(II) sulphate. This solution required 17.1 cm^3 of 0.0100 mol dm^{-3} potassium manganate(VII) solution for complete oxidation. Find the percentage iron in the sample.

The chlorine has been **reduced** because its oxidation number has **decreased**; the bromide ion has been **oxidised** because the oxidation number of bromine has increased.

To see which part of a compound is negative and which is positive in finding oxidation numbers, the **electronegativity** is used. The more electronegative atom has the negative oxidation number. So in ammonia, NH_3, nitrogen is the more electronegative and has oxidation number (-3); in nitrite, NO_2^-, it is ($+3$), and in nitrate, NO_3^-, ($+5$). Carbon in carbon dioxide, CO_2 is ($+4$), but in methane, CH_4, it is (-4). This is because carbon is less electronegative than oxygen but more electronegative than hydrogen.

Rarely, combined hydrogen does not have oxidation number ($+1$). In sodium hydride, NaH, it is combined with a less electronegative atom, and so hydrogen has oxidation number (-1) in ionic hydrides. Oxygen shows positive oxidation numbers only when combined with fluorine, e.g. it is ($+2$) in oxygen difluoride OF_2. In peroxides, like H_2O_2, we have O(-1).

Compounds or ions which apparently show fractional oxidation numbers usually have atoms of the same type with two or more different oxidation numbers. In tri-iron tetroxide, Fe_3O_4, for example, there is one Fe($+2$) and two Fe($+3$), the oxide behaving as $FeO.Fe_2O_3$. In trilead tetroxide, Pb_3O_4, the compound behaves as $2PbO.PbO_2$, i.e. as Pb($+2$) and Pb($+4$).

Using oxidation numbers to balance difficult equations

There are two ways to approach this problem. Let us apply them to the oxidation of iron(II) by manganate(VII). We shall dispense with the state symbols; they can be added at the end if required. We write the oxidation number under each element.

Method 1

$$MnO_4^- \rightarrow Mn^{2+} \qquad Fe^{2+} \rightarrow Fe^{3+}$$

$$+7 \qquad\qquad +2 \qquad +2 \qquad\qquad +3$$

$$\longrightarrow -5 \longrightarrow \qquad \longrightarrow +1 \longrightarrow$$

The -5 and $+1$ are the changes in oxidation number and they must balance (because they are just a book-keeping exercise in counting electrons). Thus $5Fe^{2+}$ must be oxidised by $1MnO_4^-$. This may be all that we need to know if we are doing a titration.

If we need to complete the equation, we know that the **stoichiometric ratio** (the relative numbers in the equation) of MnO_4^- to Fe^{2+} is 1:5. Begin by writing these down with the products:

$$MnO_4^- + 5Fe^{2+} \rightarrow Mn^{2+} + 5Fe^{3+}$$

Usually, the reactions are carried out under acidic conditions and H^+ is added to the (left-hand) side where oxygen is present and it is converted into water on the other side. Here, 4 O-atoms require $8H^+$:

$$MnO_4^- + 5Fe^{2+} + 8H^+ \rightarrow Mn^{2+} + 5Fe^{3+} + 4H_2O$$

and that's all there is to it! Of course you should check that the charges balance, as well as the atoms of each element. Here the charge is $17+$ on each side. If you have worked systematically, they will balance.

If the equation is to be used as the basis of a calculation, there is little point in hiding the important numbers (the stoichiometry) with a clutter of state symbols. Include state symbols, however, if a change of state is important to your discussion.

Method 2

Write two half-equations: one for the oxidising agent in which you add a number of electrons equal to the change in oxidation number, and enough protons (H^+) to balance the oxygen:

$$MnO_4^- + 8H^+ + 5e^- \rightarrow Mn^{2+} + 4H_2O$$

and one half-reaction representing oxidation, in which the reducing agent gives up electrons equal in number to the change in oxidation number:

$$Fe^{2+} \rightarrow Fe^{3+} + e^-$$

The transferred electrons must balance, so we need five of the second equation in this case:

$$5Fe^{2+} \rightarrow 5Fe^{3+} + 5e^-$$

When we add these two equations, the electrons cancel and we get the same equation as by Method 1:

$$MnO_4^- + 5Fe^{2+} + 8H^+ \rightarrow Mn^{2+} + 5Fe^{3+} + 4H_2O$$

This is hardly surprising since we did the same operations in a slightly different order. No-one will ask you to use a particular method; only perhaps to write an equation.

Now let us do a really nasty example with oxygen 'all over the place'!

Let us oxidise (sodium) sulphite to (sodium) sulphate using acidified potassium dichromate. I have been deliberately careless in not giving these names any oxidation numbers. In most cases, that is the name you will find on the bottle. Besides, you probably need the practice in finding oxidation numbers.

First, let us find the oxidation numbers we require.

Sulphite is SO_3^{2-}. If the oxidation number of sulphur is s, then s + 3(–2) = –2 (the charge on the ion). s = +4 (remember to include the sign) and sulphite is sulphate(IV). Similarly in sulphates, the oxidation number is +6; I leave that to you to show.

Dichromate is slightly more difficult. If the oxidation number of chromium is c, then in $Cr_2O_7^{2-}$, 2c + 7(–2) = –2; hence c = +6. It is dichromate(VI).

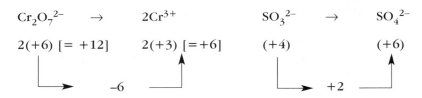

In order to balance the (oxidation number) books, one $Cr_2O_7^{2-}$ must oxidise three SO_3^{2-}:

$$Cr_2O_7^{2-} + 3SO_3^{2-} \rightarrow 2Cr^{3+} + 3SO_4^{2-}$$

On the left-hand side there are 7+(3×3) = 16 O-atoms and on the right 3×4 = 12. The difference of four O-atoms will require $8H^+$ to give $4H_2O$:

$$Cr_2O_7^{2-} + 3SO_3^{2-} + 8H^+ \rightarrow 2Cr^{3+} + 3SO_4^{2-} + 4H_2O$$

If you prefer Method 2, then try it for yourself.

Electrode potentials

Zinc metal and iron(II) salts are good reducing agents. Aqueous chlorine and potassium manganate(VII) are good oxidising agents. But how good? Chemists have a scale which measures the power or potential of an aqueous species to oxidise or reduce. It is called the electrode potential, redox potential or reduction potential scale. To understand the scale we have to examine oxidation and reduction in a little more depth. At A-level you will be expected to apply the scale; what follows may help you to understand what you are doing.

A reducing agent acts by donating electrons to the oxidising agent it reduces. Thus zinc will reduce aqueous copper ions to give copper:

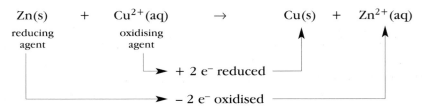

By losing two electrons, the zinc atom (the reducing agent) has become a zinc ion which, because it can take up two electrons, is an oxidising agent. Similarly

QUESTION

Deduce an equation for the oxidation of Sn^{2+} to Sn^{4+} using acidified MnO_4^- which is reduced to Mn^{2+}

the copper ion, having gained two electrons, has become a copper atom, a reducing agent.

$$Zn(s) \rightleftharpoons Zn^{2+}(aq) + 2\,e^-$$

reduced form | oxidised form
can reduce things | can oxidise things

$$Cu^{2+}(aq) + 2e^- \rightleftharpoons Cu(s)$$

oxidised form | reduced form
can oxidise things | can reduce things

We may not think of copper metal as being a reducing agent or zinc ions as an oxidising agent but that is what they are. We can choose examples which show them in this capacity.

$$Zn^{2+}(aq) + Mg(s) \rightleftharpoons Mg^{2+}(aq) + Zn(s)$$

$$Cu(s) + 2Ag^+(aq) \rightleftharpoons 2Ag(s) + Cu^{2+}(aq)$$

For every reducing agent there is an oxidised form which is a potential oxidising agent. For every oxidising agent there is a reduced form which is a potential reducing agent. We tend to think of only one of each pair because the other is not very good at doing the opposite. For example, chlorine is a strong and useful oxidising agent but its reduced form is very ineffective as a reducing agent. The situation is very similar to the Lowry-Bronsted theory of acids and bases where a conjugate pair are related by the gain or loss of a proton (H^+). Just as a strong acid has a weak conjugate base so a strong oxidising agent gives a weak reducing agent when it takes up electrons.

We are now ready to look at the chemist's list of electrode potentials and see what it means (Figure 1.1).

The scale is based on the tendency of the oxidised form of a redox pair to take up electrons forming the reduced form. The more positive the value (on the scale) the more likely this is to happen. We could imagine a system in which the reducing agent and its oxidised form were present in solution. Let us use $Fe^{2+}(aq)$ (reduced form) and $Fe^{3+}(aq)$ (oxidised form). The equation linking the species is called a half-equation and from now on we shall represent all such redox changes as reversible reductions.

$$Fe^{3+}(aq) + e^- \rightleftharpoons Fe^{2+}(aq)$$

This does not mean that $Fe^{3+}(aq)$ is a good oxidising agent and that $Fe^{2+}(aq)$ is a poor reducing agent. It is simply a standard form that we shall use. It would be very convenient if we could insert an electrical conductor – say a platinum plate which is inert and would not react with anything present – and then somehow measure the pressure of electrons from the system using some sort of "magic voltmeter" (Figure 1.2).

We might imagine that if $Fe^{2+}(aq)$ was better at reducing than $Fe^{3+}(aq)$ was at oxidising, then $Fe^{2+}(aq)$ would deposit electrons on the plate as it changed into $Fe^{3+}(aq)$, giving a negative reading on the magic voltmeter. Similarly, if $Fe^{3+}(aq)$ was a very strong oxidising agent it would take electrons from the plate as it turned into $Fe^{2+}(aq)$, giving a positive reading on the voltmeter.

REDOX EQUILIBRIA

	REDUCTION HALF-EQUATION		
	OXIDISED FORM + ne^- ⇌ REDUCED FORM		$E^{\ominus}$/V

OXIDISED FORM IS A POOR OXIDISING AGENT	Reduction half-equation		$E^{\ominus}$/V
	$Li^+(aq) + e^-$	⇌ $Li(s)$	−3.03
	$Rb^+(aq) + e^-$	⇌ $Rb(s)$	−2.93
	$K^+(aq) + e^-$	⇌ $K(s)$	−2.92
	$Ca^{2+}(aq) + 2e^-$	⇌ $Ca(s)$	−2.87
	$Na^+(aq) + e^-$	⇌ $Na(s)$	−2.71
	$Mg^{2+}(aq) + 2e^-$	⇌ $Mg(s)$	−2.37
	$Al^{3+}(aq) + 3e^-$	⇌ $Al(s)$	−1.66
	$Zn^{2+}(aq) + 2e^-$	⇌ $Zn(s)$	−0.76
	$Fe^{2+}(aq) + 2e^-$	⇌ $Fe(s)$	−0.44
	$PbSO_4(s) + 2e^-$	⇌ $Pb(s) + SO_4^{2-}(aq)$	−0.36
	$V^{3+}(aq) + e^-$	⇌ $V^{2+}(aq)$	−0.26
	$Ni^{2+}(aq) + 2e^-$	⇌ $Ni(s)$	−0.25
	$Pb^{2+}(aq) + 2e^-$	⇌ $Pb(s)$	−0.13
	$2H^+(aq) + 2e^-$	**⇌ $H_2(g)$**	**0.00**
	$S_4O_6^{2-}(aq) + 2e^-$	⇌ $2S_2O_3^{2-}(aq)$	+0.09
	$SO_4^{2-}(aq) + 4H^+(aq) + 4e^-$	⇌ $H_2SO_3(aq) + H_2O(l)$	+0.17
	$Cu^{2+}(aq) + 2e^-$	⇌ $Cu(s)$	+0.34
	$VO^{2+}(aq) + 2H^+(aq) + e^-$	⇌ $V^{3+}(aq) + H_2O(l)$	+0.34
	$IO^-(aq) + H_2O(l) + e^-$	⇌ $I^-(aq) + 2OH^-(aq)$	+0.49
	$I_2(aq) + 2e^-$	⇌ $2I^-(aq)$	+0.54
	$Fe^{3+}(aq) + e^-$	⇌ $Fe^{2+}(aq)$	+0.77
	$Ag^+(aq) + e^-$	⇌ $Ag(s)$	+0.80
	$ClO^-(aq) + H_2O(l) + e^-$	⇌ $Cl^-(aq) + 2OH^-(aq)$	+0.89
	$NO_3^-(aq) + 3H^+(aq) + 2e^-$	⇌ $HNO_2(aq) + H_2O(l)$	+0.94
	$VO_2^+(aq) + 2H^+(aq) + e^-$	⇌ $VO^{2+}(aq) + H_2O(l)$	+1.00
	$Br_2(aq) + 2e^-$	⇌ $2Br^-(aq)$	+1.09
	$MnO_2(s) + 4H^+(aq) + 2e^-$	⇌ $Mn^{2+}(aq) + 2H_2O(l)$	+1.23
	$Cr_2O_7^{2-}(aq) + 14H^+(aq) + 6e^-$	⇌ $2Cr^{3+}(aq) + 7H_2O(l)$	+1.33
	$Cl_2(aq) + 2e^-$	⇌ $2Cl^-(aq)$	+1.36
	$Mn^{3+}(aq) + e^-$	⇌ $Mn^{2+}(aq)$	+1.51
	$MnO_4^-(aq) + 8H^+(aq) + 5e^-$	⇌ $Mn^{2+}(aq) + 4H_2O(l)$	+1.51
	$MnO_4^-(aq) + 4H^+(aq) + 3e^-$	⇌ $MnO_2(s) + 2H_2O(l)$	+1.70
	$F_2(g) + 2e^-$	⇌ $2F^-(aq)$	+2.87
OXIDISED FORM IS A STRONG OXIDISING AGENT	$F_2(g) + 2H^+(aq) + 2e^-$	⇌ $2HF(aq)$	+3.06

Left margin (top to bottom): OXIDISED FORM IS A POOR OXIDISING AGENT … OXIDISING POWER OF … OXIDISED FORM … OXIDISED FORM IS A STRONG OXIDISING AGENT

Right margin (top to bottom): REDUCED FORM IS A STRONG REDUCING AGENT … REDUCING POWER OF … REDUCED FORM … REDUCED FORM IS A POOR REDUCING AGENT

Fig. 1.1 Table of standard electrode potentials.

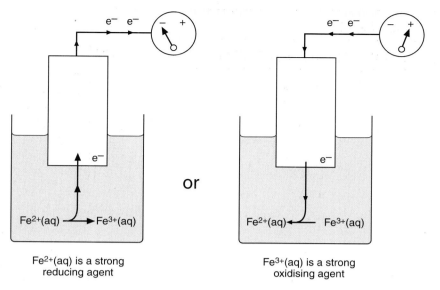

Fe²⁺(aq) is a strong
reducing agent

Fe³⁺(aq) is a strong
oxidising agent

Fig. 1.2 The magic voltmeter.

The *idea* is fine but needs improving. "Magic voltmeters" are not a recognised piece of laboratory equipment.

In order to standardise our "measurements", both oxidised and reduced forms must be present at the molar concentrations in the reduction half-equation – usually 1.0 mol dm⁻³. It would hardly be a fair test of oxidising and reducing power if only the oxidised form were present. Gases must be present at a pressure of 1.0 bar (10^5 Pa, near enough 1 atm) and the temperature must be specified; usually it is 298K. The plate making contact with the solution must be platinum if the reduced form of the system we are studying is not a conducting metal (Figure 1.3).

Fig. 1.3 Three standard electrodes (schematic).

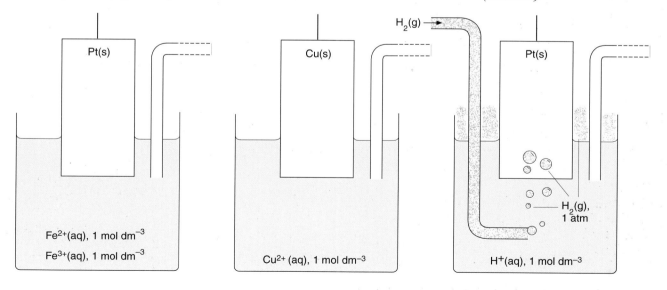

Such an arrangement is called a standard electrode. We cannot, however, devise a magic voltmeter. Electron pressure is not just like water pressure at the end of a pipe. We need an electric circuit if we are to use a voltmeter. As soon as we complete the circuit by putting a second platinum plate in the mixture we shall get another, equal and opposite electron pressure on the voltmeter. The instrument will register 0.00 V.

Fig. 1.4 Not the solution to the problem.

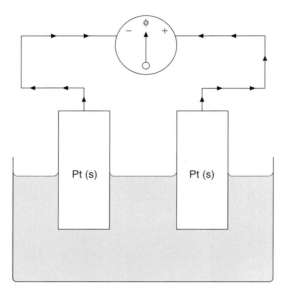

The only way we can measure this pressure is to play one electrode off against another. The connection between the two solutions must not be made using metals. A salt bridge is used. This is a conducting solution of potassium chloride in a jelly, or for crude work, in a wet filter paper bridge.

Such an arrangement can pump electrons from the more negative electrode to the less negative one and could light a suitable bulb if this replaced the

Fig. 1.5 Playing one electrode off against another.

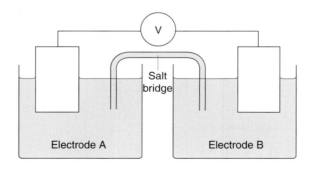

voltmeter. It is a simple cell. The voltmeter will not measure either of the 'electrode potentials' but it will measure their difference. To arrive at a value for different electrodes, one of them must be given an arbitrary value. The electrode chosen is the standard hydrogen electrode. It is conventional to show this, when present, as the left-hand component or half-cell. The half-equation is

$$H^+(aq) \; + \; e^- \; \rightleftharpoons \; \tfrac{1}{2} \, H_2(g)$$

The standard electrode potential is 0.00 V. This is written after the half-equation; the superscript $\ominus$ indicates standard conditions

$$H^+(aq) \; + \; e^- \; \rightleftharpoons \; \tfrac{1}{2} \, H_2(g) \quad E^\ominus = 0.00 \text{ V}$$

To find any electrode potential, the electrode in question (usually standard) is made to be the right-hand component of a cell in which the left-hand component is a standard hydrogen electrode. The e.m.f. (reading on a high-resistance voltmeter) of the cell is, by convention, given the sign of the right-hand electrode.

If, for example, the electrode was a zinc plate in 1.0 mol dm^{-3} Zn^{2+}(aq) (e.g.

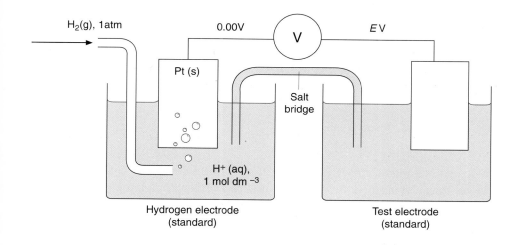

Fig. 1.6 *Arrangement for finding electrode potentials (schematic).*

aqueous zinc sulphate) the voltmeter would show –0.76 V (zinc negative). We would write:

$$Zn^{2+}(aq) \; + \; 2e^- \; \rightleftharpoons \; Zn(s) \quad E = -0.76 \text{ V}$$

Another, more convenient way is to use a subscript showing the oxidised and reduced forms in that order

$$E^\ominus_{Zn^{2+}|Zn} \; = \; -0.76 \text{ V}$$

Electrode potential may thus be defined as the e.m.f. of a cell in which the left-hand component is a standard hydrogen electrode and the right-hand component is the electrode system in question. If the standard electrode potential is required, then the conditions in the test electrode must also be standard.

You are not required to know how to set up such a cell. Hydrogen electrodes are difficult to use and, in practice, they are often replaced with other standard electrodes of known potential. All the values for common systems are known and tabulated. You are required to predict the feasibility of changes in a way which shows that you understand the principles.

Electrode potentials and the feasibility of reactions

The feasibility of redox reactions in aqueous solution is assessed by comparing electrode potentials. It is important that you do this in a special way.

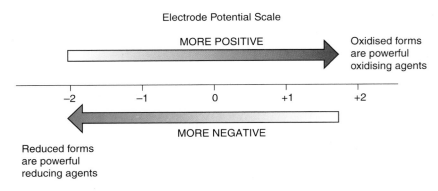

Fig. 1.7 *Playing the 'numbers game'.*

When comparing electrode potentials there is one rule: to say 'bigger' or 'larger than', 'smaller' or 'less than' is incorrect. When comparing numbers you must say that one is 'more negative' or 'more positive' than the other. Always include the sign of numbers, positive or negative. Thus

–2 is more negative than –1

–1 is more negative than +1

+2 is more negative than +3

+3 is more positive than +1

Remember that all half-equations are written as reductions. One will eventually have to be reversed, as oxidation and reduction occur simultaneously but, at the start, all are written as reductions. Let us look at a familiar situation. Will zinc metal reduce copper(II) ions or will copper metal reduce zinc ions?

$$Zn^{2+} \text{ (aq)} + 2\,e^- \rightleftharpoons Zn \text{ (s)} \qquad E^\circ = -0.76 \text{ V}$$

$$Cu^{2+} \text{ (aq)} + 2\,e^- \rightleftharpoons Cu \text{ (s)} \qquad E^\circ = +0.34 \text{ V}$$

This example is particularly easy (especially as we already know what happens!). The first half-equation represents a negative electrode (potential) and the second a positive one. The $Zn^{2+}|Zn$ electrode potential is more negative, so the first equation provides the electrons (by going backwards), and the second equation uses the electrons to reduce Cu^{2+}.

From the more negative zinc system, the reduced form, Zn, provides the electrons:

$$Zn\ (s)\ \rightarrow\ Zn^{2+}\ (aq)\ +\ 2e^-$$

the half-equation is reversed to provide electrons; and from the more positive system, the oxidised form, Cu^{2+}, takes up the electrons:

$$Cu^{2+}(aq)\ +\ 2\,e^-\ \rightarrow\ Cu(s)$$

The equation is thus:

$$Zn\ (s)\ +\ Cu^{2+}\ (aq)\ \rightarrow\ Cu\ (s)\ +\ Zn^{2+}\ (aq)$$

By writing any two half-equations as reductions and comparing their electrode potentials, we can predict the feasibility of a redox reaction in solution.

> The half-equation with the more negative reduction potential will reveal the reducing agent. The half-equation will go backwards (make negative progress).

> The half-equation with the more positive reduction potential will reveal the oxidising agent. The half-equation will go forwards (make positive progress).

'Make positive progress' and 'make negative progress' are aids to memory which might give you confidence in your answer. You must not treat them as explanations – they are not acceptable on an exam paper. We can safely say that:

> The probable direction of reaction is that in which the reduced form of the system (electrode) with the more negative electrode potential will reduce the oxidised form of the system with the more positive electrode potential.

Let us look at two examples where E has the same sign for both electrodes.

Example 1
If you do not like $\frac{1}{2}$ in an equation, you can of course double both half-equations; if you do the value of E remains the same. E is like temperature; it doesn't matter how much of the North Sea you swim in, it's just as cold.

REDOX EQUILIBRIA

$$\tfrac{1}{2} \, Cl_2 \, (aq) \ + \ e^- \ \rightarrow \ Cl^- \, (aq) \qquad E^\ominus = + \, 1.36 \ V$$

$$\tfrac{1}{2} \, I_2 \, (aq) \ + \ e^- \ \rightarrow \ I^- \, (aq) \qquad E^\ominus = + \, 0.54 \ V$$

QUESTION

If the Daniell cell,
$Zn(s) \, | \, Zn^{2+}(aq) \ \rightleftharpoons$
$Cu^{2+}(aq) \, | \, Cu(s)$, is set up
under standard conditions
and an opposing potential
greater than 1.10 V is applied
to the terminals of the cell,
what changes would you
expect?

E for the first equation is more positive, so that the oxidised form in that half-equation is going to oxidise (draw electrons from) the reduced form in the other.

The equations are written as reductions, so if the chlorine is to accept electrons (be reduced) the direction is correct (more positive E gives positive progress). The more negative iodine reduction potential means that the (reduced form in the) second equation provides the electrons, i.e. the I^- provides them when the equation has been reversed (more negative E gives negative progress):

$$\tfrac{1}{2} \, Cl_2(aq) \ + \ I^- \, (aq) \ \rightarrow \ \tfrac{1}{2} I_2(aq) \ + \ Cl^-(aq)$$

Now let us take two negative E values and try to solve a problem to which we do not know the answer.

Example 2

Will a piece of manganese and a tin(II) chloride solution react to give tin and a manganese(II) solution? The two half-equations for the reduction of manganese(II) and tin(II) are:

$$Mn^{2+}(aq) \ + \ 2e^- \ \rightarrow \ Mn(s) \qquad E^\ominus = -1.19 \ V$$

$$Sn^{2+}(aq) \ + \ 2e^- \ \rightarrow \ Sn(s) \qquad E^\ominus = -0.14 \ V$$

E for the Mn half-equation is the more negative and it is the electron provider. That can only happen if it is reversed

$$Mn \, (s) \ \rightarrow \ Mn^{2+} \, (aq) \ + \ 2e^- \quad (oxidation)$$

whereas the more positive E-value for the tin system shows that it will absorb electrons, as written

$$Sn^{2+} \, (aq) \ + \ 2e^- \ \rightarrow \ Sn \, (s)$$

On adding

$$Mn \, (s) \ + \ Sn^{2+} \, (aq) \ \rightarrow \ Sn(s) \ + \ Mn^{2+} \, (aq)$$

The reaction as predicted is feasible. But will it happen?

When does a feasible reaction not happen?

There are three common reasons that prevent a feasible reaction from occurring.

1. The reaction takes a more favoured course than the one we thought of.

Suppose that we wish to know whether manganate(VII) ions will oxidise iron(II). We consult tables of electrode potentials and find:

$$MnO_4^-(aq) + 4H^+(aq) + 3e^- \rightarrow MnO_2(s) + 2H_2O(l) \quad E^\ominus = +1.70 \text{ V}$$

$$Fe^{3+}(aq) + e^- \rightarrow Fe^{2+}(aq) \quad E^\ominus = +0.77 \text{ V}$$

E for the first equation is more positive, therefore (the oxidised form in) the first equation will take electrons from the (reduced form in the) second equation, resulting in oxidation. The more positive MnO_4^- equation goes forward and the more negative Fe^{3+} equation is reversed (to provide electrons) and multiplied by 3 (to provide the right number):

$$MnO_4^-(aq) + 4H^+(aq) + 3e^- \rightarrow MnO_2(s) + 2H_2O(l)$$

$$3Fe^{2+}(aq) \rightarrow 3Fe^{3+}(aq) + 3e^-$$

giving

$$MnO_4^-(aq) + 4H^+(aq) + 3Fe^{2+}(aq) \rightarrow MnO_2(s) + 2H_2O(l) + 3Fe^{3+}(aq)$$

However, in volumetric analysis we rely on 1 mol of MnO_4^- (aq) oxidising 5 mol of Fe^{2+} (aq) to give a clear and almost colourless solution. The above equation, leading to an ugly brown precipitate soon after the titration begins, often occurs when the analyst forgets to add sulphuric acid at the start. If we compare the equation with that of the reaction more familiar to us,

$$MnO_4^-(aq) + 8H^+(aq) + 5Fe^{2+}(aq) \rightarrow Mn^{2+}(aq) + 4H_2O(l) + 5Fe^{3+}(aq)$$

we see that this alternative reaction is preferred provided that plenty of acid is present. Yes, oxidation occurred, but our choice of reduction product was the wrong one.

> **QUESTION**
> State a practical objection to the use of the (standard) hydrogen electrode.

2. The activation energy of the reaction is too high.

If we add zinc to dilute hydrochloric acid there is a slow but steady evolution of hydrogen.

$$Zn^{2+}(aq) + 2e^- \rightleftharpoons Zn(s) \quad E^\ominus = -0.76 \text{ V}$$

$$H_2(g) + 2e^- \rightleftharpoons 2H^+(aq) \quad E^\ominus = 0.00 \text{ V}$$

The $Zn^{2+}|Zn$ system has the more negative reduction potential and Zn will thus be the electron provider, reducing H^+ to hydrogen.

If we add copper to a dilute acid then use the electrode potential for the $Cu^{2+}|Cu$ system

$$Cu^{2+}(aq) + 2e^- \rightleftharpoons Cu(s) \qquad\qquad E^\ominus = +0.34\ V$$

$$H_2(g) + 2e^- \rightleftharpoons 2H^+(aq) \qquad\qquad E^\ominus = 0.00\ V$$

we see that copper has the more positive system and its oxidised form, Cu^{2+}, is the only possible reactant. Copper should not react with dilute acids to give hydrogen; however, hydrogen (E more negative) should apparently reduce a solution of copper sulphate to give copper. If you bubble hydrogen gas through such a solution you will be disappointed. The covalent H–H would have to be broken which requires a large activation energy, and hydrogen is almost completely insoluble in water as well.

3. The values of E are very close and/or something escapes from the system.

All redox reactions reach equilibrium but the position of equilibrium is usually so far to the left or right that this can be ignored. This does not happen when the reduction potentials are very close together (see below) unless something 'escapes' (forms a gas or a complex etc.) from the system and disturbs the equilibrium. It accounts for the tendency of some reactions to go in unfavourable directions.

Disproportionation

This occurs when, in the same reaction, the oxidation number of an element is both increased and decreased (in different products). Explaining it with E values is no more difficult than our earlier examples provided that you have the relevant half-equations. However, students do find difficulty in writing the appropriate half-equations.

If you warm a little (red) copper(I) oxide with dilute sulphuric acid you will obtain a brown precipitate of copper and a blue solution of copper(II) sulphate. The copper(I) ions disproportionate

$$2\ Cu^+(aq) \rightarrow Cu(s) + Cu^{2+}(aq)$$

oxidation number $\qquad 2\times(+1) \qquad\quad 0 \qquad\qquad +2$

Notice that there is no overall increase or decrease. There has been no external oxidising or reducing agent. This redox change is entirely a property of the copper(I) ion.

To explain it we must first write two reductions (even though we know that one must end as an oxidation) and look up their electrode potentials. In one copper(I) must be reduced and in the other it must be produced by reduction.

$$Cu^+(aq) + e^- \rightarrow Cu(s) \qquad\qquad E^\ominus = +0.53\ V$$

$$Cu^{2+}(aq) + e^- \rightarrow Cu^+(aq) \qquad\qquad E^\ominus = +0.15\ V$$

Notice that the equation

$$Cu^{2+} \text{ (aq)} + 2\,e^- \rightleftharpoons Cu \text{ (s)} \qquad E^\ominus = +0.34 \text{ V}$$

which we have used earlier, is no use to us. This is one of the problems you may find when trying to deal with this topic. The last equation does not contain the species Cu^+ (aq) (which is disproportionating) but it does contain the other two. Our equations, I repeat, must both contain the disproportionating species, once *being reduced* and once as a *reduction product*.

QUESTION

Find and use appropriate electrode potentials to show whether disproportionation will occur in the following cases under standard conditions:

(a) $VO^{2+} \rightarrow VO_2^+$ and V^{3+}

(b) $MnO_2 \rightarrow MnO_4^-$ and Mn^{2+}

Returning to the two relevant equations we see that the upper one has the more positive electrode potential and (the oxidised form in it) will thus draw electrons from (cause oxidation in) the lower equation, which, as it represents a reduction, will have to be reversed. The driving force or cell potential being $0.53 - 0.15 = 0.38$ V. We can add this information to our original equation

$$2\,Cu^+ \text{ (aq)} \rightarrow Cu \text{ (s)} + Cu^{2+} \text{ (aq)} \qquad E^\ominus_{cell} = +0.38 \text{ V}$$

Our argument supports our findings that disproportionation occurs. In general, if $E^\ominus_{cell}$ is positive the change from left to right is favoured (see p. 17).

Let us see if an iron(II) salt might be expected to disproportionate into iron and an iron(III) salt.

$$3\,Fe^{2+} \text{ (aq)} \rightleftharpoons 2\,Fe^{3+} \text{ (aq)} + Fe(s)$$

You might like to try to decide without looking at the solution below.

QUESTION

Write an ionic equation for the disproportionation of Sn^{2+}(aq) into tin and Sn^{4+}(aq). Use electrode potentials to predict whether it is likely to happen.

The relevant equations are

$$Fe^{2+} \text{ (aq)} + 2\,e^- \rightleftharpoons Fe \text{ (s)} \qquad E^\ominus = -0.41 \text{ V}$$

$$Fe^{3+} \text{ (aq)} + e^- \rightleftharpoons Fe^{2+} \text{ (aq)} \qquad E^\ominus = +0.77 \text{ V}$$

The second equation has the more positive electrode potential and thus the reaction will go forward attracting electrons from the first reaction, which, having the more negative E value, will provide electrons by being reversed.

$$2\,Fe^{3+} \text{ (aq)} + 2\,e^- \rightleftharpoons 2\,Fe^{2+} \text{ (aq)}$$

$$Fe \text{ (s)} \rightleftharpoons Fe^{2+} \text{ (aq)} + 2\,e^-$$

On adding these we get the equation

$$2\,Fe^{3+} \text{ (aq)} + Fe(s) \rightleftharpoons 3\,Fe^{2+} \text{ (aq)}$$

This equation is exactly the opposite of what we were looking for and shows that, far from iron(II) salts disproportionating, an aqueous iron(III) salt would react with iron to produce an iron(II) salt. The 'driving force' or cell potential is the difference between the two E values, in this case

$$E^\ominus_{cell} = 0.77 - (-0.41) = +1.28 \text{ V}$$

Non-standard conditions

We have discussed all these reactions using E values for standard conditions. The conditions we choose are most unlikely to be standard. How much difference does it make?

Temperature and concentration changes are likely to affect the rate of a reaction rather than its direction. The effect, for ionic reactions in solution, which are very fast, is seldom noticeable. The most significant effect of concentration is usually observed with acids. Small changes of concentration hardly affect E values but the changes of concentration of H_3O^+ with pH are enormous. The number of possibilities is vast and the best advice is: "If the conditions are non-standard – beware!".

Cells

A cell is a device that, by chemical action, produces an electric current. We saw how electrode potentials are found by constructing a cell. In principle a cell can be made by opposing any two electrode systems of different electrode potential. The electrode with more negative electrode potential provides electrons (as its reduced form changes into its oxidised form); it is thus the negative pole of the cell. Similarly the electrode with more positive electrode potential is the positive pole.

In the cell a redox reaction occurs. Unlike a simple reduction-oxidation in a mixed solution, the electrons used are not passed 'uselessly' from the reductant to the oxidant. Instead, the reduction of the oxidising agent and the oxidation of the reducing agent occur in separate half-cells and the electrons pass (current) through an external wire, under the pressure difference (voltage) caused by the difference in the electrode potentials of the two half-cells, whence they can do useful work.

One of the first useful cells was made in the early nineteenth century by J. F. Daniell, long before the nature of atoms and ions or the existence of the electron was known. It consisted of a zinc plate in aqueous zinc sulphate separated from a copper plate in aqueous copper(II) sulphate by a porous pot (Figure 1.8).

Although it was widely used for fixed installations, such as the early telegraph, its large size, low voltage and, above all, its inconvenient liquid content soon rendered it obsolete.

Representing electrodes and cells

The simplest electrodes are represented by the metal's symbol (the reduced form of the electrode) followed by a vertical line, which indicates a phase change, (s) to (aq), and then the solution. Components of the solution are separated by a comma. The two electrodes in the Daniell cell are shown as

$$\text{Zn (s)} \mid \text{Zn}^{2+}\text{(aq, 1 mol dm}^{-3}) \quad \text{and} \quad \text{Cu (s)} \mid \text{Cu}^{2+}\text{(aq, 1 mol dm}^{-3})$$

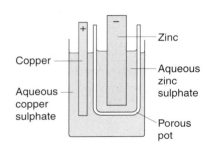

Copper

Aqueous copper sulphate

Zinc

Aqueous zinc sulphate

Porous pot

Fig. 1.8 A Daniell cell.

When the two electrodes are opposed in a cell the two are separated by two vertical dotted lines, ⫶, representing the salt bridge. Because they are opposed, the right-hand half-cell is reversed.

$$\text{Zn (s)} \mid \text{Zn}^{2+}\text{(aq, 1 mol dm}^{-3}) \; ⫶ \; \text{Cu}^{2+} \text{(aq, 1 mol dm}^{-3}) \mid \text{Cu (s)}$$

this is often simplified to

$$\text{Zn (s)} \mid \text{Zn}^{2+}\text{(aq)} \; ⫶ \; \text{Cu}^{2+} \text{(aq)} \mid \text{Cu (s)}$$

A convention that we shall adopt is to give the e.m.f. of such a cell (the voltage or p.d. between the electrodes measured with a very high resistance voltmeter) the sign of the right-hand electrode, copper in this cell. From this, it follows that

$$E_{cell} = E_{\text{RH electrode}} - E_{\text{LH electrode}}$$

Thus the e.m.f. of the standard Daniell cell (ideally) is $+0.34 - (-0.76)$ $= + 1.1$ V.

QUESTION

How would you represent a cell with zinc and silver electrodes?

The advantage of such a convention is that if we make a habit of arranging cells so that they have a positive e.m.f., then the right-hand electrode is always taking electrons from the left-hand one and we know that oxidation is happening at the left-hand electrode. Likewise reduction is happening at the right (positive) one. There is one situation where we deliberately do not do this: when we are using a hydrogen electrode it is always written on the left.

There is a further advantage of the convention in which the cells are arranged as shown,

Reduced form of A,	Oxidised form of A	⫶	Oxidised form of B,	Reduced form of B

QUESTION

How would you expect the e.m.f. of a Daniell cell to alter (qualitatively) if 0.1 mol dm^{-3} ZnSO$_4$ and 2.0 mol dm^{-3} CuSO$_4$ were used (in the same cell) instead of 1.0 mol dm^{-3} solutions?

(where the comma may be replaced by a phase-change line if the reduced form is a metal). If the e.m.f of the cell is positive, the changes in both electrodes are from left to right (forwards); if it is negative, the reverse is true. Thus, when the cell above (with zinc and copper electrodes) is working (i.e. current is flowing), the changes are as indicated.

$$\text{Zn(s)} \mid \text{Zn}^{2+}\text{(aq)} \; ⫶ \; \text{Cu}^{2+}\text{(aq)} \mid \text{Cu(s)}$$

This can be used as a rather mechanical way of solving redox problems using electrode potentials. It is, in principle, exactly what we have done before. You have to remember that if you reverse a reduction half-equation you must reverse the sign of the electrode potential, but if you multiply the equation you do not alter the magnitude of E.

Example: Will cobalt reduce gallium(III) salts in solution to gallium?

$$3\text{Co(s)} + 2\text{Ga}^{3+}\text{(aq)} \rightarrow 3\text{Co}^{2+}\text{(aq)} + 2\text{Ga(s)}$$

The two reduction half-equations are:

$$Co^{2+}(aq) + 2e^- \rightarrow Co(s) \qquad E^\ominus = -0.28\,V$$
$$Ga^{3+}(aq) + 3e^- \rightarrow Ga(s) \qquad E^\ominus = -0.56\,V$$

Rearranging and multiplying to get the desired equation:

$$3Co(s) \rightarrow 3Co^{2+}(aq) + 6e^- \qquad E^\ominus = +0.28\,V$$
$$2Ga^{3+}(aq) + 6e^- \rightarrow 2Ga(s) \qquad E^\ominus = -0.56\,V$$

$$\overline{3Co(s) + 2Ga^{3+}(aq) \rightarrow 3Co^{2+}(aq) + 2Ga(s) \qquad E^\ominus = +0.28 + (-0.56)\,V}$$

$$E^\ominus_{cell} \text{ (or } E^\ominus_{overall}) = -0.28V$$

As the value of $E^\ominus_{cell}$ for the reaction as written is negative, the *reverse* reaction will be favoured, *not* the one proposed.

Practical cells and batteries (storage cells)

Cells or 'batteries' (strictly a collection of cells) fall into two groups: rechargeable and disposable. The rechargeable cells (e.g. lead-acid or nickel-cadmium) can have the redox processes which gave rise to the electricity reversed by 'electrolysis' – passing current in the reverse direction. The greater complexity of the rechargeable cells means that they are either much bigger than their throw-away counterparts or else they are discharged more quickly. On the whole, the rechargeable cells tend to have lower voltages.

Simple 'batteries' used in radios, torches, watches and hearing aids tend to have chemically complicated electrodes in which the bulk content is in the form of a stiff paste for safety in transport and use.

Titrations

Potassium manganate(VII) titrations

Potassium manganate(VII) is an intensely purple substance, and gives a pink colour even in very dilute solution. It is its own indicator. It is usually used in the burette, so that the slight excess present in the reaction mixture when the endpoint is reached is easily seen as a pink coloration. The half-reaction for its reduction in acidic solution, the usual conditions of use, is:

$$MnO_4^-(aq) + 8H^+(aq) + 5e^- \rightarrow Mn^{2+}(aq) + 4H_2O(l)$$

Potassium manganate(VII) is a strong oxidising agent and cannot be obtained in a high state of purity. It easily oxidises organic materials, say specks of dust present in the solutions, and these slowly deposit manganese(IV) oxide on standing. For these reasons, solutions cannot be made accurately by weighing the solid, so they must be standardised against a solution which can be so made (a primary standard). Sodium ethanedioate $Na_2C_2O_4$ is commonly used as a standard. Ammonium iron(II) sulphate is good enough for elementary work, but the solution must be prepared within a day or two of use.

QUESTION

A quantity of sodium ethanedioate was weighed out and dissolved in water to make 250 cm^3 of solution. A 25.0 cm^3 portion was acidified with dilute sulphuric acid, warmed to 60°C and titrated with 0.0200 mol dm^{-3} potassium manganate(VII) solution; 27.3 cm^3 was required for complete oxidation. How much sodium ethanedioate was originally weighed out, assuming it to have been pure?

The reaction for the oxidation of ethanedioate ions is:

$$C_2O_4^{2-}(aq) \rightarrow 2CO_2(g) + 2e^-$$

hence

$$2MnO_4^-(aq) + 5C_2O_4^{2-}(aq) + 16H^+(aq)$$
$$\rightarrow 2Mn^{2+}(aq) + 10CO_2(g) + 8H_2O(l)$$

The potassium manganate(VII) solution is run into a standard solution of sodium ethanedioate at about 60°C; initially the reaction is quite slow, but as soon as some manganese(II) ion is produced this catalyses the reaction which becomes much faster. Such a reaction where the products catalyse the reaction is called **autocatalytic**.

Reference to tables of electrode (or redox) potentials shows that MnO_4^-/H^+ is a very powerful oxidising agent and will oxidise almost any reducing agent which requires titration.

Sodium thiosulphate titrations

Sodium thiosulphate is used to titrate iodine. The thiosulphate is oxidised to tetrathionate ions, $S_4O_6^{2-}$, while the iodine forms iodide ions:

$$2S_2O_3^{2-}(aq) + I_2(aq) \rightarrow S_4O_6^{2-}(aq) + 2I^-(aq)$$

The substance to be analysed usually produces the iodine in another reaction by oxidation of iodide ions. Thus aqueous copper(II) will oxidise iodide ions:

$$2Cu^{2+}(aq) + 4I^-(aq) \rightarrow 2CuI(s) + I_2(aq)$$

so the liberated iodine can be titrated with sodium thiosulphate solution. This titration can be used to determine the amount of copper in brass, for example, or the number of molecules of water of crystallisation in hydrated copper(II) sulphate.

As titrations of iodine with thiosulphate ions proceed, the solution becomes pale yellow; at the endpoint it turns colourless (unless other coloured substances are present), but since this can be difficult to see, freshly prepared starch solution is often added near the endpoint. The solution turns deep blue-black due to the formation of a starch/iodine complex, which disappears at the endpoint.

Sodium thiosulphate is oxidised to sulphate ions by other halogens, but is not used for their analysis.

Corrosion

Corrosion is the conversion of a metal in its normal working environment, mostly iron in practice, to its ions. It is therefore oxidation, and many circumstances where corrosion occurs involve electrochemical cells.

REDOX EQUILIBRIA

Fig. 1.9 Rusting – the remains of a car in an Australian scrapyard.

Rusting

Rust is a hydrated iron oxide, $Fe_2O_3.xH_2O$, and is commercially one of the most significant compounds of iron. Enormous amounts of money are involved in preventing rusting and in replacing manufactured goods that have fallen victim to it.

Rusting requires oxygen and a film of *liquid* water on the iron. Water vapour is not enough, since the water film forms the electrolyte in which corrosion occurs.

Attack on the iron occurs where there are traces of impurities or points of strain; these cause minute variations in the electrode potential of iron. Water must be present. Oxygen, however, is not necessary at the point of attack; that is why pitting often occurs on car bodies under loosely adhering paint work where water can make its way by capillary attraction from breaks in the paint. The iron dissolves to give a limited concentration of iron(II) ions (Fig. 1.10a):

$$Fe(s) \rightleftharpoons Fe^{2+}(aq) + 2e^-$$

Fig. 1.10a Rusting – the initial attack.

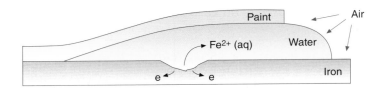

The released electrons pass through the iron to some point where water and oxygen are present. The electrons reduce this mixture to hydroxide ions (Fig. 10b):

$$2 H_2O (l) + O_2 (g) + 4 e^- \rightleftharpoons 4 OH^- (aq)$$

Fig. 1.10b Rusting – initial action of oxygen.

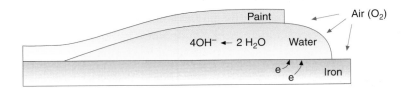

Any acidic gases in the water, e.g. carbon dioxide, assist this reaction by removal of the hydroxide ions

$$OH^- (aq) + CO_2 (g) \rightleftharpoons HCO_3^- (aq)$$

Rusting is notably faster in industrial areas.

The iron(II) ions are oxidised to iron(III) when they come into contact with oxygen (often at the exposed metal where the water entered under the paint):

$$4\,Fe^{2+}\,(aq)\ +\ 2\,H_2O\,(l)\ +\ O_2\,(g)\ \rightleftharpoons\ 4\,Fe^{3+}\,(aq)\ +\ 4\,OH^-\,(aq)$$

In the oxygenated region, the iron(III) and hydroxide ions react to deposit brown iron(III) hydroxide:

$$Fe^{3+}\,(aq)\ +\ 3\,OH^-\,(aq)\ \rightleftharpoons\ Fe(OH)_3\,(s)$$

which 'ages' to give rust

$$2\,Fe(OH)_3\,(s)\ \rightleftharpoons\ \text{'}Fe_2O_3.H_2O\,(s)\text{'}$$

which is a mixture of hydrates of iron(III) oxide (Fig. 1.10c).

Fig. 1.10c Deposition of rust.

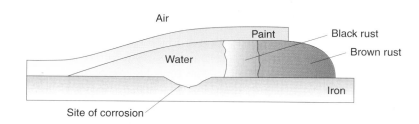

The rust is porous and allows further entry of water. Where the air is restricted, for example behind the rust under the paint, the hydroxide ions react with iron(II) ions which have not yet been oxidised:

$$Fe^{2+}\,(aq)\ +\ 2\,OH^-\,(aq)\ \rightleftharpoons\ Fe(OH)_2\,(s)$$

This results in a mixture of iron oxides in different oxidation states which is the cause of black rust.

If you precipitate a little iron(II) hydroxide by adding excess aqueous sodium hydroxide to a dilute solution of iron(II) sulphate and continuously shake it with air, you will see the precipitate go greener, then go almost black before it turns brown. Alternatively – for quick results – add the aqueous sodium hydroxide to a mixture of iron(II) and iron(III) salts in solution.

Corrosion of other metals

Although rusting is quantitatively the most significant form of corrosion, aluminium and magnesium alloys used in aircraft and ships will corrode. Corrosion is often found where two different metals are in contact, for example aluminium alloys riveted with magnesium alloy rivets, which forms a magnesium–aluminium cell ($E^{\ominus} = +0.71V$), or steel bolts used with alloy components. Such corrosion is kept at bay by suitable maintenance and protection using various water-repellent materials as well as frequent cleaning and painting.

Fig. 1.11 Tinning, the process of coating steel cans with tin, prevents corrosion and poses no risk to health.

Sacrificial protection

Electrochemistry can be exploited in the prevention of rusting, as well as being the cause of it. A more reactive metal in contact with iron will corrode preferentially. This is used in galvanising, where the object is dipped in molten zinc. If the zinc coating is damaged, a cell is formed and the zinc will oxidise rather than the iron.

Iron ships and underground pipes, both occupying wet and perhaps salty environments, are protected by blocks of zinc or, more usually, magnesium, attached at intervals to the steel. These blocks corrode instead of the steel, and are replaced when necessary.

In these cases the intent is that the more reactive metal is sacrificed to preserve the steel, hence its being called a sacrificial coating.

Tinning

Zinc cannot be used to protect steel items used to contain foodstuffs. Although needed in small quantities for health, large amounts of zinc are toxic. Steel is therefore coated with tin instead (Figure 1.11). Tin is less easily oxidised than iron ($E^{\ominus}_{Sn^{2+}|Sn} = -0.14V$) so it does not act as a sacrificial coat if it is damaged. Under such conditions, the iron does corrode. The purpose of the tin is simply to protect the iron with a tough, malleable, non-toxic coating. The preservation of food in 'tins' marked a dramatic advance in public health and food safety in the nineteenth century.

Questions

1 (a) Give the formula of an oxoanion in which

   (i)   N has an oxidation number of $+5$

   (ii) N has an oxidation number of $+3$

   (iii) Cl has an oxidation number of $+1$

   (iv) Cl has an oxidation number of $+5$

   (v)   Al has an oxidation number of $+3$

   (b) Write a balanced equation in which the ion in (iii) disproportionates into chloride ions and the ion in (iv).

2 State what, if anything, has been oxidised or reduced in each of the following changes. Justify your answers by calculating relevant oxidation numbers.

   (i) $H_2 + S \rightarrow H_2S$

   (ii) $2\,Al + 3\,Cl_2 \rightarrow Al_2Cl_6$

   (iii) $2\,OF_2 \rightarrow O_2 + 2\,F_2$

   (iv) $S + SO_3^{2-} \rightarrow S_2O_3^{2-}$

   (v) $H_2O + SO_3 \rightarrow 2\,H^+ + SO_4^{2-}$

(vi) $H^+ + SO_4^{2-} \rightarrow HSO_4^-$

(vii) $SO_3 + H_2SO_4 \rightarrow H_2S_2O_7$

(viii) $2\,CrO_4^{2-} + 2\,H^+ \rightarrow Cr_2O_7^{2-} + H_2O$

3 Predict, using values from Fig. 1.1 (p. 6) whether:

(i) metallic lead would reduce aqueous copper(II) sulphate to copper

(ii) acidified nitrates would oxidise aqueous sodium chloride to chlorine

(iii) acidified dichromate(VI) would oxidise manganese(IV) oxide to manganate(VII)

(iv) aqueous VO^{2+} would be reduced to aqueous V^{3+} (VO^+) by metallic silver

(v) an aqueous solution of vanadium in the +4 oxidation state would be oxidised to the +5 oxidation state by chlorine

(vi) an aqueous solution of vanadium in the +4 oxidation state would disproportionate to a mixture of compounds of vanadium in the +3 and +5 oxidation states.

4 A cell is made using a magnesium rod in aqueous magnesium sulphate (1.0 mol dm^{-3}) and a nickel rod in aqueous nickel(II) sulphate (1.0 mol dm^{-3}).

(i) Write a symbolic representation of the cell (starting with the magnesium half-cell)

(ii) Calculate the e.m.f. of the cell and state which rod is the positive terminal

(iii) Write an equation for the overall chemical reaction when the cell provides external current

(iv) Suggest, with your reason, which half-cell might be replaced by copper in copper(II) sulphate in order to increase the e.m.f. of the cell.

5 A 25.0 cm^3 sample of dilute hydrochloric acid is treated with potassium iodate(V) and potassium iodide in sufficient quantity to liberate iodine and use up all the hydrogen ions available. The iodine was titrated with sodium thiosulphate solution of concentration 0.100 mol dm^{-3}, 27.8 cm^3 being required. Find the concentration in mol dm^{-3} of the hydrochloric acid.

6 A sample of a copper alloy weighing 2.83 g was treated so as to convert it to 250 cm^3 of a neutral solution containing copper as copper(II) ions. 25.0 cm^3 portions of this solution were treated with excess potassium iodide, and the liberated iodine titrated with 0.100 mol dm^{-3} sodium thiosulphate solution. 26.7 cm^3 was required. Find the percentage by mass of the copper in the alloy.

Transition metals

The d-block and transition metals

The d-block of the Periodic Table is best defined by shading it in on the Table! It is not easy to describe briefly. It includes the ten elements numbered 21 (Sc) to 30 (Zn) and the elements arranged directly below them. The block is so called because (i) an atom of the first member of each horizontal row (or period) of the block is the first to contain one electron in the d orbitals of a specified subshell, (ii) an atom of the tenth element in each horizontal row has ten electrons in those d orbitals, and (iii) the electronic structure of an atom of the element following that last element is made by adding an electron to the first of the p orbitals of the outer (valency) shell.

A d-block element is described as a transition element if it can form an ion, stable under normal laboratory conditions, with a partially filled set of d orbitals. Elements outside the d-block never form ions with a partially filled set of d-orbitals in normal reactions.

The aufbau principle (Unit 1) shows that the 3d subshell fills after the 4s, which is complete at element 20, calcium $1s^2 2s^2 2p^6 3s^2 3p^6 4s^2$ (or $[Ar]4s^2$). Scandium is therefore $[Ar]3d^1 4s^2$, and zinc at the other end of the series is $[Ar]3d^{10} 4s^2$. Neither scandium nor zinc are transition metals since the only ion scandium forms is Sc^{3+}, with no d electrons, and zinc forms only Zn^{2+} ($3d^{10}$), with a full d subshell.

Because the orbitals being filled are inner ones, the change in chemistry across the series is less marked than the change which is seen across a similar number of elements in the main groups, where electrons are being added to the outermost shell. In transition metals the increase in nuclear charge is offset to a large extent by increased screening of the outer electrons by the inner electrons, so ionisation energies do not vary very much, nor do atomic radii. Vertical trends are much less significant than those in main groups. The second and third transition series are left to more advanced work.

Electronic structures and variable oxidation numbers

First series transition metal electronic structures are given in Table 2.1, the 3d and 4s orbitals being used. As we go from left to right in the d-block, the pattern of adding one electron to the inner shell each time we increase the atomic number, with two electrons in the outer shell, is not maintained. There are energy advantages in having the d orbitals half-full or full, and this is the reason for the unexpected structures for Cr and Cu.

Table 2.1 *Electron configurations in the d-block*

	Sc	Ti	V	Cr	Mn	Fe	Co	Ni	Cu	Zn
3d	1	2	3	5	5	6	7	8	10	10
4s	2	2	2	1	2	2	2	2	1	2

A paired electron is usually more easily removed from a given orbital than an unpaired electron; remember the kink in the graph of first ionisation energies, Na–Ar, at S and P. This sort of energy difference helps to explain why Fe^{2+} (d^6) easily forms Fe^{3+} (d^5) but Mn^{2+} (d^5) is relatively difficult to oxidise ($E^{\ominus}_{Fe^{3+}|Fe^{2+}} = +0.77$ V; $E^{\ominus}_{Mn^{3+}|Mn^{2+}} = +1.51$ V).

Owing to increasing nuclear charge, the atomic radii fall from calcium to chromium. Further addition of electrons to the 3d shell then makes relatively little difference, with size rising slowly because of increased screening of the outer electrons from the nucleus, until there is a small rise at zinc. Figure 2.1 shows the metallic radii.

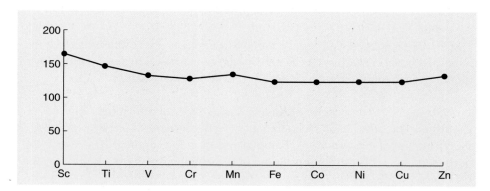

Fig.2.1   Metallic radii/pm, for the first transition series.

The same gradual trend is seen for first and second ionisation energies (Figure 2.2), in complete contrast to the trends in main group elements.

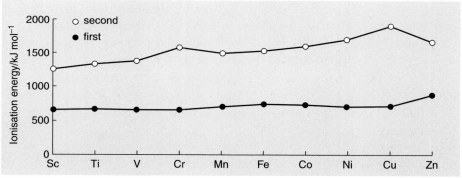

Fig. 2.2   First and second ionisation energies in the first transition series.

Successive ionisation energies for a given atom rise by a similar amount each time, until there is a large jump when all 4s and 3d electrons have been removed. Figure 2.3 shows successive ionisation energies for vanadium [Ar] $3d^34s^2$ and for copper [Ar] $3d^{10}4s^1$.

Fig. 2.3 Successive ionisation energies for vanadium and copper.

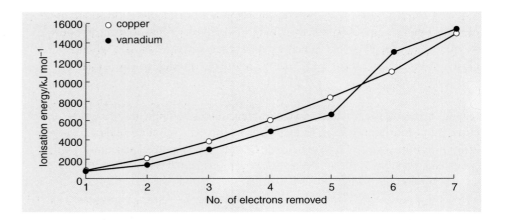

The increases between successive ionisation energies are compensated for by increased bond strengths, e.g. in oxoanions, or increased hydration enthalpy of the ions, enabling the transition metal to have several oxidation numbers.

Variable valency is one of the distinguishing features of transition metal chemistry. The other main characteristics are the formation of complex ions, the formation of coloured ions, paramagnetism in ions and catalytic activity in both elements and compounds. Some of these will now be considered further.

Formation of complex ions

Complex ions consist of a central metal 'ion' surrounded by neutral molecules, e.g. H_2O or NH_3, or anions, e.g. Cl^- or CN^-, called **ligands.** The ligands always have lone pairs of electrons which they use to form dative covalent bonds (coordinate bonds) with the central ion. There are commonly six ligands and six coordinate bonds arranged octahedrally (Figure 2.4).

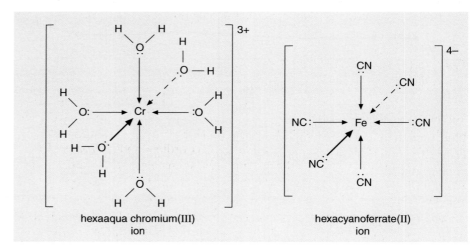

hexaaqua chromium(III) ion

hexacyanoferrate(II) ion

Fig. 2.4 Two octahedral complex ions (the darker and broken arrows denote in front and behind).

The overall charge on the complex ion is the sum of the charges on the central cation and the ligands. If the complex is a cation, it is named by using a prefix for the ligands followed by the name of the central ion with its oxidation number; e.g. hexaaquairon(II), $[Fe(H_2O)_6]^{2+}$, or hexaamminenickel(II), $[Ni(NH_3)_6]^{2+}$. If adding negatively charged ligands has caused an overall negative charge and the complex is an anion, the name of the central ion is changed to ferrate (Fe), cuprate (Cu) or something ending in –ate which is recognisable from the symbol for the central atom; e.g., hexacyanoferrate(II), $[Fe(CN)_6]^{4-}$. Notice that the Roman numeral denotes the (original) oxidation number of the central cation not the charge on the complex ion.

The complexes owe their stability (relative to the ions or molecules from which they are made) to the loss of energy when lone pairs, confined to an atom or ion (the ligand) are shared, forming a bond. Some ligands bond more strongly than others. Water (aqua-) is quite a weak ligand compared with, say, ammonia (ammine-) or cyanide ion (cyano-). The addition of the ligand (or its compound, e.g. K^+CN^-) to an aqueous solution of a transition metal ion often replaces the aqua ligand:

$$[Fe(H_2O)_6]^{2+} \ + \ 6CN^- \ \rightleftharpoons \ [Fe(CN)_6]^{4-} \ + \ 6H_2O$$

The bonding of cyanide ions with iron(II) ions is so strong that a 'fast-and-furious' first-aid remedy for cyanide poisoning is to force the victim to swallow a freshly prepared suspension of iron(II) hydroxide (prepared with sodium carbonate, not sodium hydroxide). The free cyanide ions are deadly; if complexed in time, they represent less of a threat.

Complex formation is not limited to d-block ions. Their small radii, large positive charge and possession of available valency shell orbitals makes the formation of coordinate bonds particularly favourable. Magnesium and beryllium, in the s-block, are notable for their ability to complex. The striking thing about the transition metal ions is the colour of their complexes.

Complex formation and colour

You saw in Unit 1 that the d (atomic) orbitals fall into two groups. Two of the five orbitals ($d_{x^2-y^2}$ and d_{z^2}) have axes (regions of highest electron density) which correspond to the direction of the x, y and z axes of a three-dimensional graph. The other three orbitals lie between these axes. When six ligands approach to form an octahedral complex, the coordinate bonds (pairs of electrons) are directed down the x, y and z axes. Any electrons in the two d orbitals of the central cation already centred on these axes will be repelled by the incoming pairs. This raises the energy of the electrons in the d orbitals – like compressing a spring. The three 'between axis' d orbitals are less affected. Thus the energy of the two on-axis orbitals is raised above the other three (Figure 2.5).

The difference in energy, ΔE, corresponds to the absorption (or emission) of electromagnetic radiation which lies in the visible spectrum:

$$\Delta E = h\upsilon \quad \text{or} \quad hf \ (\upsilon \text{ or } f = \text{frequency})$$

TRANSITION METALS

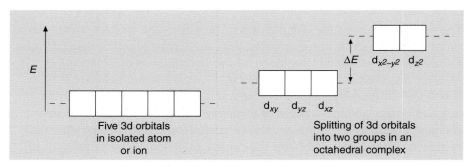

Fig. 2.5 *Effect of ligands on d orbitals.*

This light absorption is possible only if the d orbitals are neither completely full (when there are no vacant orbitals to which an electron can be promoted) nor completely empty (when there are no electrons to promote). Thus transition metal complexes are normally coloured but some elements at the extreme ends of the d block do not form coloured complexes. In Sc^{3+} (and Ti^{4+}) the d orbitals are empty. Cu^+ and Zn^{2+} (which readily forms complexes) have full d-shells, [Ar] $3d^{10}$, and their complexes are colourless. Anhydrous copper(II) sulphate is colourless because it is not an octahedral complex. However, aqueous copper(II) sulphate is blue because it has a partially filled d-shell *and* is an octahedral aqua-complex (Figure 2.6).

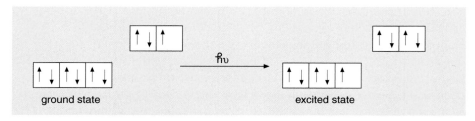

Fig. 2.6 *Absorption of light in aqueous copper(II) sulphate.*

Strictly, if anything is covalently linked to a cation (or central atom) it forms a complex ion. Thus all the oxoanions of the d-block are, technically, complex. They are often intensely coloured; e.g. manganate(VII) (or tetraoxo-manganate(VII), if you wish to draw attention to its complex nature). They are not octahedral and the cause of the colour is less simply explained.

Atomic orbitals, e.g. 1s, 3p, in isolated atoms or ions, have precisely defined energy levels. The light absorbed when electrons are promoted, or emitted as they fall back to a lower energy level, as in the 'flame test', is of a precise frequency. This gives a sharp line in the spectrum.

The difference in energy of the d orbitals in a complex depends on the extent to which the orbitals are compressed by the ligand electric field. The geometry of a complex alters slightly because of thermal vibration, hence (ΔE and υ are not precise, they vary from one ion to the next. The light absorption is over a wide range of frequencies. You will meet this idea again when we look at ultraviolet spectra later in this book.

When the d orbital energies are altered by complex formation, any property which depends on them e.g. the removal or addition of electrons, will also be affected. For this reason, the electrode potential of 'simple' aqueous solutions of transition metal ions is not the same as that of other complexes of the same ions in the same oxidation states. In the pair below, quoted as an example, the two 'simple' species Fe^{2+}(aq) and Fe^{3+}(aq) should, strictly, be quoted as the aqua complexes, $[Fe(H_2O)_6]^{2+}$(aq) and $[Fe(H_2O)_6]^{3+}$(aq), but this is not normally done in tables of reduction potentials:

$$Fe^{3+}(aq) \ + \ e^- \ \rightleftharpoons \ Fe^{2+}(aq) \quad E^{\ominus} = \ +0.77 \text{ V}$$

$$[Fe(CN)_6]^{3-}(aq) \ + \ e^- \ \rightleftharpoons \ [Fe(CN)_6]^{4-}(aq) \quad E^{\ominus} = \ +0.36 \text{ V}$$

QUESTION
Why are scandium(III) compounds colourless?

The action of alkali on aqua complexes

Solutions of sodium hydroxide or ammonia are used in the qualitative analysis of transition metal ions, and the reactions illustrate some of their properties. Aqua ions of transition metals are acidic. The water ligands are strongly polarised by the electron-withdrawing effect of the small highly-charged metal ion, so protons are removed in a nucleophilic attack by the solvent water. For hexaquairon(III) the following reaction takes place:

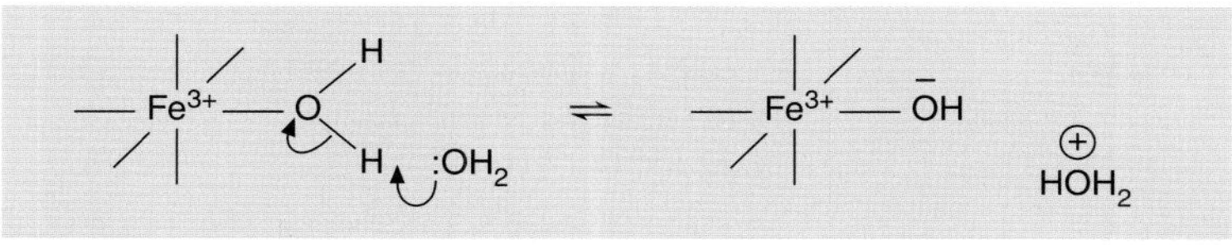

This equilibrium can be moved further to the right by adding hydroxide ions, which are much more basic than water and so therefore much better at deprotonating the complex. The result is that the metal hydroxide precipitates:

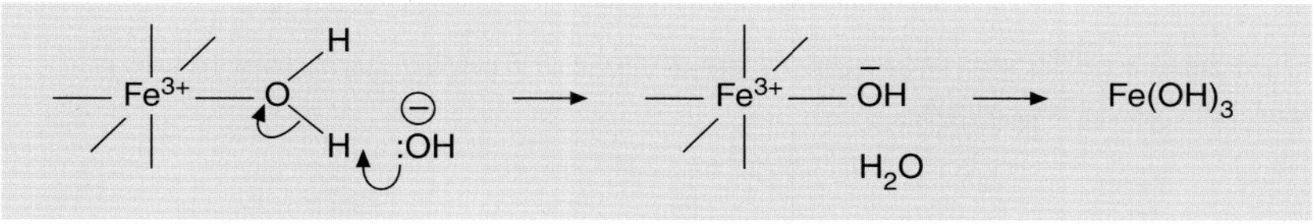

The colours of the precipitates (Table 2.2) can be used to identify the ions. The reaction is the same whether the hydroxide ion comes from sodium hydroxide or from ammonia solution via the reaction

$$NH_3(aq) + H_2O(l) \ \rightleftharpoons \ NH_4^+(aq) + OH^-(aq)$$

In this reaction water is acting as an acid, donating the protons to the ammonia. Whether water acts as acid or base depends on the acid or base strength of the other substance.

Many transition metal ions react further if more alkali is added. The reactions are often different for sodium hydroxide and ammonia. Since sodium hydroxide is a strong base, further reaction depends on the other metal hydroxide also having acidic properties as well as its expected basic ones, i.e. it must be amphoteric. Chromium(III) hydroxide is an example. It reacts as a base with aqueous acid

$$Cr(OH)_3(s) + 3H^+(aq) \rightarrow Cr^{3+}(aq) + 3H_2O(l)$$

QUESTION
Suppose you have solutions of an iron(II) salt and an iron(III) salt containing $[Fe(H_2O)_6]^{2+}$ and $[Fe(H_2O)_6]^{3+}$, of equal molar concentration. Which would be more acidic, and why?

TRANSITION METALS

Table 2.2 *Reactions of transition metal ions with sodium hydroxide and ammonia. All reactions are in aqueous solution, so (aq) is omitted. The transition metal ions are hexaaqua species $[M(H_2O)_6]^{n+}$*

Ion	With NaOH or NH₃ not in excess	With NaOH in excess	With NH₃ in excess
Ti^{4+}	$Ti^{4+} + 4OH^- \rightarrow Ti(OH)_4$. White precipitate	No further reaction	No further reaction
V^{2+}	$V^{2+} + 2OH^- \rightarrow V(OH)_2$. Violet precipitate	No further reaction	No further reaction
V^{3+}	$2V^{3+} + 6OH^- \rightarrow V_2O_3 + 3H_2O$. Green precipitate	No further reaction	No further reaction
Cr^{3+}	$Cr^{3+} + 3OH^- \rightarrow Cr(OH)_3$. Green or violet precipitate	$Cr(OH)_3 + 3OH^- \rightarrow [Cr(OH)_6]^{3-}$. Deep green solution	No further reaction
Mn^{2+}	$Mn^{2+} + 2OH^- \rightarrow Mn(OH)_2$. Buff precipitate which darkens in air as MnO_2 forms	No further reaction	No further reaction
Fe^{2+}	$Fe^{2+} + 2OH^- \rightarrow Fe(OH)_2$. Pale green precipitate which turns brown at surface due to oxidation to $Fe(OH)_3$ with air	No further reaction	No further reaction
Fe^{3+}	$Fe^{3+} + 3OH^- \rightarrow Fe(OH)_3$. Red-brown precipitate	No further reaction	No further reaction
Co^{2+}	$Co^{2+} + 2OH^- \rightarrow Co(OH)_2$. Blue ppt which turns brown in air as $Co(OH)_3$ is formed	No further reaction	$Co(OH)_2 + 6NH_3 \rightarrow [Co(NH_3)_6]^{2+} + 2OH^-$. Blue solutions which oxidise in air to $[Co(NH_3)_6]^{3+}$
Ni^{2+}	$Ni^{2+} + 2OH^- \rightarrow Ni(OH)_2$. Pale green precipitate	No further reaction	$Ni(OH)_2 + 6NH_3 \rightarrow [Ni(NH_3)_6]^{2+} + 2OH^-$. Pale lavender-blue solution
Cu^{2+}	$Cu^{2+} + 2OH^- \rightarrow Cu(OH)_2$. Pale blue precipitate	No further reaction, although the precipitate is appreciably soluble	$Cu(OH)_2 + 4NH_3 + 2H_2O \rightarrow [Cu(NH_3)_4(H_2O)_2]^{2+} + 2OH^-$. Deep blue solution

Note: Iron(II) hydroxide, when freshly prepared from very pure iron(II) salt solutions, in the absence of iron(III) impurities or air, is a pale bluish white. Traces of iron(III), or oxidation, result in a green precipitate.

and as an acid with aqueous base

$$Cr(OH)_3(s) + 3OH^-(aq) \rightarrow [Cr(OH)_6]^{3-}(aq)$$

The result is a deep-green solution of the ion $[Cr(OH)_6]^{3-}$. Addition of acid protonates this and it reverts to $Cr(OH)_3$ with the green precipitate reappearing.

d-Block metal hydroxides may have a variety of structures, ranging from hydrated hydroxide complexes, e.g. $[Cr(OH)_3(H_2O)_3]$ formed by loss of three protons from the hydrate, to loose associations of the metal oxide with water. Representing them by $M(OH)_n$ is simply a convenience.

Ammonia is too weak a base to cause the amphoteric hydroxides to react as acids. Some transition metals will instead form complexes with the ammonia as ligand, such complexes being soluble. Copper(II) provides a good example. On adding ammonia to an aqueous copper(II) salt, a pale blue gelatinous

> ## QUESTION
>
> Suggest how, using either sodium hydroxide solution or ammonia solution as appropriate, you could separate the ions in aqueous solutions of
> (a) Zn^{2+} and Cu^{2+};
> (b) Fe^{2+} and Cr^{3+}.

precipitate of copper(II) hydroxide, $Cu(OH)_2$, appears; with excess ammonia this reacts to give a deep blue, soluble complex. The nature of this depends rather on one's interpretation of copper chemistry. Some regard the complex as square planar with four ammonia ligands; others as octahedral with two water ligands in addition. Taking the latter view, the reaction is

$$Cu(OH)_2(H_2O)_4(s) + 4NH_3(aq) \rightarrow [Cu(NH_3)_4(H_2O)_2]^{2+}(aq) + 2OH^-(aq) + 2H_2O(l)$$

The reaction is a **ligand exchange**.

Table 2.2 summarises some reactions of ions from the first transition series. The equations should be known, together with the observations. For convenience the aqua ions are usually written, for instance, as $Cr^{3+}(aq)$ rather than $[Cr(H_2O)_6]^{3+}$, but remember that the ligands are there.

Vanadium

Vanadium is not a common metal and there are no large-scale ore deposits. However, it is useful in alloy steels, which are subject to shock and which need to be springy, for example for making spanners. Also, vanadium(V) oxide, V_2O_5, is used in the Contact Process for manufacture of sulphuric acid, as the catalyst for the oxidation of sulphur dioxide to sulphur trioxide.

It is easy to show the interconversion of the various oxidation states of vanadium, and the colours are pretty (Table 2.3).

Table 2.3 *Some compounds of vanadium*

Oxidation state	d-orbital electrons	Common ion in water	Other compounds
V(+5)	0	yellow, VO_2^+,	VF_5, VO_3^-, V_2O_5
V(+4)	1	blue, VO^{2+}	VCl_4, VO_2
V(+3)	2	green, VO^+ or $[V(H_2O)_6]^{3+}$	V_2O_3
V(+2)	3	lavender, $[V(H_2O)_6]^{2+}$	

V(+5) is mildly oxidising, V(+4) is the most stable state in presence of air, V(+3) the most stable in absence of air, and V(+2) is strongly reducing.

Ammonium metavanadate, NH_4VO_3, is a colourless solid which dissolves in acid to give a yellow solution of VO_2^+. Addition of zinc powder and shaking produces, successively, green (a mixture of VO_2^+ and VO^{2+}), blue (VO^{2+}), green (V^{3+} or VO^+) and lavender (V^{2+}).

Vanadium(V) exists only as fluorine or oxygen compounds, for instance VF_5 and V_2O_5. Vanadium(V) oxide, V_2O_5, is a brown solid, is quite acidic, and is used as a catalyst.

QUESTION

Justify the view that vanadium is a transition metal, giving examples of the characteristic properties you would expect.

Vanadium(IV) can be produced by mild reduction of vanadium(V), for example with aqueous sulphur dioxide (e.g. by using acidified sulphite solutions):

$$2VO_2^+(aq) + SO_3^{2-}(aq) + 2H^+(aq) \rightarrow 2VO^{2+}(aq) + SO_4^{2-}(aq) + H_2O(l)$$

This difficult equation can be arrived at using half-equations and oxidation numbers:

$$2VO_2^+ + 4H^+ + 2e^- \rightarrow 2VO^{2+} + 2H_2O$$
$$2\times(+5) \qquad\qquad 2\times(+4)$$

$$SO_3^{2-} + H_2O \rightarrow SO_4^{2-} + 2H^+ + 2e^-$$
$$+4 \qquad\qquad +6$$

The first half-equation has been doubled throughout so that the numbers of electrons (or the changes in oxidation number) are the same in both half-equations. The two half-equations can then be added and the state symbols added.

Or with iron(II):

$$VO_2^+(aq) + Fe^{2+}(aq) + 2H^+(aq) \rightarrow VO^{2+}(aq) + Fe^{3+}(aq) + H_2O(l)$$

Vanadium(III) needs more powerful reducing agents for its production than does vanadium(IV), and is readily oxidised by air. Further reduction to vanadium(II) is quite slow, and although a number of vanadium(II) compounds are known, the aqueous solution is strongly reducing and will even react with solvent water if no other oxidising agent is available:

$$2V^{2+}(aq) + 2H_2O(l) \rightarrow 2V^{3+}(aq) + H_2(g) + 2OH^-(aq)$$

QUESTION

Write the half-equations for the redox reaction of VO_2^+ with Fe^{3+}.

Iron

Iron is the second most abundant metal in the earth's crust (aluminium is the most abundant), and its engineering importance needs no comment here.

Iron is usually used in mild steel, an alloy with about 0.5% carbon. It oxidises very readily in the presence of water and oxygen to give the highest common oxidation state of iron as iron(III) oxide. Rust is the hydrated form of this, and it flakes off the surface exposing fresh metal to corrosion (see p.20).

Iron(III) is marginally the most stable oxidation state in air. The electron structure is $[Ar]3d^5$. Apart from rust, which commercially is the most significant iron(III) compound, the chloride, $FeCl_3$, is important. It exists in anhydrous and hydrated forms.

Anhydrous iron(III) chloride is a nearly black covalent solid. The Fe^{3+} ion is quite small and polarising enough to give a covalent compound with chlorine. Dimerisation to Fe_2Cl_6 occurs, and $FeCl_3$ can be used as a catalyst in the Friedel-Crafts reaction (Chapter 4). It is thus similar to $AlCl_3$.

Iron(III) chloride dissolves in water to give yellow solutions from which the hydrated salt $FeCl_3.6H_2O$ can be crystallised. This is used to etch 'printed' circuit boards in electronics, where unprotected copper is oxidised:

$$Cu(s) + 2Fe^{3+}(aq) \rightarrow Cu^{2+}(aq) + 2Fe^{2+}(aq)$$

If Fe^{3+} solutions are made very acidic (pH < 0), the colour changes to amethyst; this is seen nicely in the double salt ammonium iron(III) sulphate, and in amethysts themselves which consist of quartz with iron(III) impurity.

Hexaaquairon(III), $[Fe(H_2O)_6]^{3+}$, is more acidic than the hexaaquairon(II) complex. It is deprotonated by sodium hydroxide or aqueous ammonia to give a red-brown precipitate of iron(III) hydroxide, $Fe(OH)_3$, which does not react further with either reagent.

Anhydrous iron(II) chloride, $FeCl_2$, is white. With water, it gives a pale green solution of $[Fe(H_2O)_6]^{2+}$, which normally contains traces of iron(III) owing to atmospheric oxidation, although making the solution acidic slows this considerably. Addition of sodium hydroxide or aqueous ammonia precipitates dirty-green, impure iron(II) hydroxide, $Fe(OH)_2$, which does not react further with the reagents but will oxidise quite quickly in air to iron(III) hydroxide, $Fe(OH)_3$, which is red-brown.

A useful analytical test for iron ions is to use an iron complex with cyanide. For iron(III), Fe^{3+}, potassium hexacyanoferrate(II), $K_4Fe(CN)_6$, is used. For iron(II), Fe^{2+}, potassium hexacyanoferrate(III), $K_3Fe(CN)_6$, is used. In both cases you get an intense 'Prussian blue' precipitate. The test is very sensitive. Prussian blue was used by the Prussian army to colour their uniforms, for printing engineering blueprints, and for the locomotives of the Somerset & Dorset Joint Railway.

QUESTION
Write equations to show
(a) the reduction of iron(III) oxide to iron using carbon monoxide,
(b) the reduction of iron(III) oxide to iron using carbon. Both reactions occur in the blast furnace.

Catalytic activity and the transition metals

A catalyst increases the rate of a chemical reaction, but is not consumed in the process. It works by providing an alternative mechanism for the reaction to the uncatalysed one, with a lower energy barrier to be overcome. There are two types, homogeneous, where the catalyst is in the same phase as the reactants (gas, liquid, solution), and heterogeneous, where it is not. Heterogeneous catalysts are usually solids catalysing gas or liquid phase reactions.

A phase is a region of matter in which neither the physical state nor the composition varies (except at a molecular level).

Transition metal ions can function as homogeneous catalysts because they can move between different oxidation states. An example is the slow reaction between cerium(IV), Ce^{4+} and thallium(I), Tl^+:

$$2Ce^{4+}(aq) + Tl^+(aq) \rightarrow 2Ce^{3+}(aq) + Tl^{3+}(aq)$$

QUESTION
Write the equation for the oxidation of $Fe^{2+}(aq)$ to $Fe^{3+}(aq)$ using oxygen. The half-reaction for the reduction of oxygen is:

$$O_2 + 4H^+ + 4e^- \rightarrow 2H_2O$$

This is catalysed by Mn^{2+} ions; their action is thought to be due to their ability to form manganese(III) and manganese(IV). Cerium(IV) oxidises manganese(II) successively to manganese(III) and manganese(IV). Mn(IV) then oxidises the thallium(I) ion, and reverts back to manganese(II).

Heterogeneous catalysis involves the combination of reactants with the catalyst surface in some way. This weakens the bonds within the reactants, allowing them to form the products which then leave the catalyst surface. The exact sequence of events is complex, and depends on the strength of the bonds in the reactants and of those formed with the catalyst surface, as well as the size of the atoms in the lattice of the catalyst and the orbitals which it can make available to bond with the reactants. These must not bond strongly with the products. Thus a substance fulfilling all these requirements for a given reaction will not be common, and may not work for another reaction unless it is very similar. Catalysts are usually specific to particular reactions or types of reaction.

Table 2.4 gives details of some industrially important, catalysed reactions.

Table 2.4 *Industrially important catalysed reaction*

Process	Reaction	Catalyst and conditions
Haber process: synthesis of ammonia	$3H_2(g) + N_2(g) \rightleftharpoons 2NH_3(g)$	Iron with traces of potassium and aluminium oxides; 350–1000 atm, 350°C
Hydrogenation of alkenes: used in manufacture of margarine	$RCH{=}CHR' + H_2 \rightarrow RCH_2CH_2R'$	Platinum or palladium at room temperature (both very expensive); nickel at 50–150°C (used commercially)
Contact process: manufacture of sulphuric acid	$2SO_2(g) + O_2(g) \rightleftharpoons 2SO_3(g)$	Vanadium(V) oxide, V_2O_5; 1.5 atm, 400°C.
Ostwald process: oxidation of ammonia in the manufacture of nitric acid	$4NH_3(g) + 5O_2(g) \rightarrow 4NO(g) + 6H_2O(g)$ NO reacts further with air to NO_2, which then reacts with water	Platinum–rhodium alloy in the form of a gauze; 900°C
Sandmeyer reaction	$C_6H_5N_2^+ + Cl^- \rightarrow C_6H_5Cl + N_2$	Copper(I) chloride
Ziegler-Natta polymerisation of alkenes	e.g. $CH_2{=}CH_2 \rightarrow$ polymer with highly controlled structure	Complex of $TiCl_3$ and $Al(C_2H_5)_3$; various conditions depending on desired product

Questions

1   Suggest explanations for the following observations. Identify any unnamed species and explain the chemistry involved.

   (*a*)   A pale green, acidified solution of iron(II) sulphate goes golden brown on the addition of hydrogen peroxide.

   (*b*)   A yellow solution of a potassium salt goes orange on the addition of sulphuric acid and then green on adding calcium sulphite, $CaSO_3$.

   (*c*)   A pale blue solution goes deep blue on the addition of concentrated aqueous ammonia but green on the addition of concentrated hydrochloric acid.

(d) When purple crystals of potassium chromium(III) sulphate-12-water are dissolved in cold water they form a purple solution, but on leaving the solution to stand or on boiling it, the solution goes green.

2 Cobalt(II) salts catalyse the decomposition of household bleach, aqueous sodium chlorate(I).

$$2 \text{ ClO}^- \text{ (aq)} \rightarrow 2 \text{ Cl}^- \text{ (aq)} + \text{O}_2 \text{ (g)}$$

The likely catalytic activity of the cobalt ion is that it is first oxidised to the cobalt(III) state by the chlorate and then the water is oxidised to oxygen by the cobalt(III) intermediate.

(i) Give the electronic structures of the Co^{2+} and Co^{3+} ions.

(ii) State how you would identify one product of the reaction.

(iii) Explain why Ca^{2+} would be unlikely to catalyse the reaction.

(iv) Fe^{2+} ions accelerate the decomposition of chlorates but the iron is in the +3 oxidation state at the end of the reaction. State how this satisfies, or fails to satisfy, the usual definition of a catalyst.

(v) Suggest, giving your reasons, another cation which might catalyse this reaction.

(vi) Aqueous solutions of chlorate(I) are strongly alkaline.

$$\text{ClO}^- \text{ (aq)} + \text{H}_2\text{O (l)} \rightleftharpoons \text{HOCl (aq)} + \text{OH}^- \text{ (aq)}$$

In the above question we have been careful not to write equations showing the oxidation of Co^{2+}(aq) reversibly to Co^{3+}(aq). Suggest why.

3 Silver salts are usually colourless. (Silver bromide and iodide are slightly yellow because, in the solid, the silver cation distorts the orbitals of the anion. This has no bearing on the question.) Oxidation of a silver nitrate solution in pyridine, $\text{C}_5\text{H}_5\text{N}$, by a peroxodisulphate, $\text{S}_2\text{O}_8^{2-}$ (all colourless compounds) produced bright orange crystals of a silver(II) complex, $[\text{Ag}(\text{C}_5\text{H}_5\text{N})_4]^{2+}\text{S}_2\text{O}_8^{2-}$. The structure of pyridine differs from that of benzene in just one way; one C–H in the benzene ring is replaced in pyridine by N: (with an unshared pair of electrons). The electron structure of silver is $(\text{Kr})4\text{d}^{10}5\text{s}^1$. Suggest

(i) why silver salts are normally colourless, but the complex was orange.

(ii) how the pyridine molecules were held to the central silver ion.

3 Benzene – the basis of aromatic compounds

The study of organic chemistry is divided into two main sections:
- **aliphatic compounds**; those compounds based on carbon chains,
- **aromatic compounds**; those based on a benzene ring structure.

The compounds studied in previous units have been aliphatic. This chapter introduces the study of aromatic compounds. In order to understand the nature of aromatic compounds, it is first necessary to understand the structure of benzene itself.

Structure of benzene

The problem: The structure of benzene proved to be a great problem to chemists for many years because:
- the molecular formula is C_6H_6. This can easily be shown from the percentage composition by mass, determined experimentally .
- the benzene molecule has a regular hexagonal shape and is planar. This can be shown from a variety of modern techniques such as X-ray diffraction (see Figure 3.4).
- the C–C bond length in benzene is 0.139 nm which is roughly half-way between the C–C bond length in ethane (0.154 nm) and that of C=C in ethene (0.134 nm).
- benzene undergoes substitution reactions rather than addition.

The solution: From the knowledge and experience gained in previous units, we would expect a molecule such as this to be highly unsaturated and to contain numerous double, or even triple, covalent bonds. One such structure could be

$$CH_2=C=CH–CH=C=CH_2$$

Structures such as this, however, would be expected to undergo numerous addition reactions, but this is not the case for benzene.

The Kekulé approach

The first breakthrough in our understanding of the structure came from a German chemist, Friedrich August Kekulé (1829–96) (Figure 3.1). He proposed a structure in which the carbon atoms form a regular hexagon. His structure was:

Fig. 3.1 Friedrich August Kekulé (1829–96), who first proposed a structure for benzene.

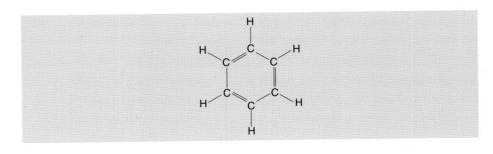

which is represented in the skeletal form as:

This structure looks unsaturated. The shape of the molecule would not be a regular hexagon and the bond lengths would not all be equal. It would be expected to undergo addition reactions in the same way as ethene since it appears to have three localised pi bonds. Indeed it might be called cyclohexatriene.

The problem can be resolved by representing the structure as a hybrid of two forms:

Fig. 3.2 A model of the arrangement of the sigma bonds in benzene.

The double-headed arrow is taken to mean that the actual structure lies somewhere between the two representations. Each C–C bond is therefore neither a single nor a double bond, but is of an intermediate type lying between the two forms. Benzene is then described as being a **resonance hybrid** of the two structures. This does NOT mean that the structure oscillates between the two structures but rather that benzene has a single structure in which the C–C bonds are all of the same nature, this being something in between a double bond and a single bond. This concept overcomes the difficulties regarding the shape and the bond lengths of the benzene molecule.

The molecular orbital approach

The molecular orbital approach is based on the carbon atoms being sp^2 hybridised, with three planar orbitals at angles of 120° to one another. Linear overlap between these orbitals could therefore lead to a planar hexagonal structure in which the carbon atoms are all linked by sigma bonds (Figure 3.2). Similarly, overlap with the 1s orbitals of the six hydrogen atoms would form six more sigma bonds which would also be in the same plane. Each carbon atom still has a p orbital containing a single electron and these protrude above and below the plane of the carbon ring . Each of these p orbitals overlaps laterally with the two p orbitals on either side of it forming a pi molecular orbital. This molecular orbital is different from those seen in alkenes, however, since it is not formed by just two atomic orbitals but by six. The molecular orbital is therefore not in one position between two particular C atoms but is spread all round the carbon ring and is said to be **delocalised**. A model of benzene based on this approach to the structure is shown in Figure 3.3.

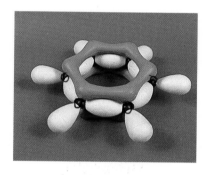

Fig. 3.3 A model of the structure of the benzene ring showing the delocalised ≠-bond

An electron density contour map, obtained by X-ray diffraction, is shown in Figure 3.4 and this clearly demonstrates that the high electron density is spread equally around the whole ring and not concentrated in particular areas.

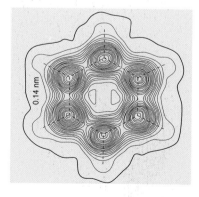

Fig. 3.4 An electron density contour map of benzene obtained by X-ray diffraction.

BENZENE – THE BASIS OF AROMATIC COMPOUNDS

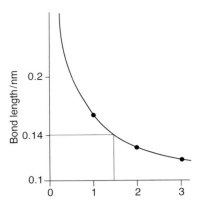

Fig. 3.5 *Bond order in benzene.*

If we plot the known carbon–carbon bond lengths in ethane (0.154 nm), ethene (0.134 nm) and ethyne (0.121 nm) against the bond order (C–C, 1), (C=C, 2) and (C≡C, 3), we can use the observed bond length in benzene (Figure 3.5) to determine the bond order. As the graph shows, this is about 1.5. This is just what we would expect with 18 electrons being shared between six equal bonds.

You may represent benzene in structures, equations and mechanisms as either the fixed bond (Kekulé) or delocalised bond structure. You must *not* use the cyclohexane structure.

Thermochemical evidence for the structure of benzene

Cyclohexatriene does not exist. We cannot do any experiments on it. All we can do to arrive at its thermochemical properties is to calculate them. We do this by looking at related structures (Method 1) or by using bond enthalpies (Method 2).

Method 1: Comparing enthalpy changes of hydrogenation

It is not necessary to determine these enthalpy changes by experiment. This is fortunate, because hydrogenation does not lend itself to such measurements. The values can be safely calculated from easily measured enthalpy changes of combustion:

$$\Delta H_{hydrogenation}$$

$$X \ + \ H_2(g) \longrightarrow XH_2$$

$$\Delta H_c(X) \ + \ \Delta H_c(H_2) \qquad \Delta H_c(XH_2)$$

combustion products of X and H_2O

You can work out the formula linking these enthalpy changes yourself.

It is reasonable to assume that the enthalpy change on hydrogenation of cyclohexene

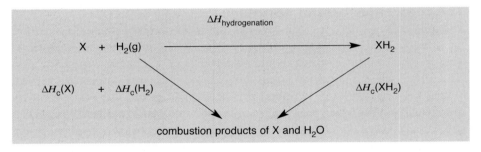

$$+ \ H_2 \longrightarrow \qquad \Delta H^\ominus = -120 \, \text{kJ} \, \text{mol}^{-1}$$

will be one-third of that for cyclohexatriene.

$$+ \ 3H_2 \longrightarrow \qquad \Delta H^\ominus = 3 \times (-120) = -360 \, \text{kJ} \, \text{mol}^{-1}$$

We cannot test this, of course, because we cannot make this compound. However, we can make cyclohexa-1,3-diene, and we would expect its enthalpy change of hydrogenation to be $2 \times (-120) = -240$ kJ mol^{-1}. Thermochemical measurements show that it is sufficiently close to this figure, at $\Delta H_h = -233$ kJ mol^{-1}, to allow us to proceed. (The discrepancy of 7 kJ is because the molecule is stabilised by limited delocalisation of the conjugated double bonds.)

When we examine the corresponding enthalpy change for benzene, however:

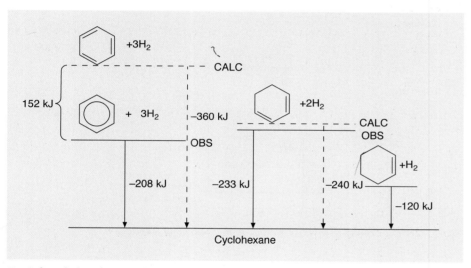

we see that it cannot be cyclohexatriene. The stabilising effect of the delocalisation is an enormous $360 - 208 = 152$ kJ mol^{-1} (Figure 3.6).

Fig. 3.6 Enthalpy changes of hydrogenation.

Method 2: Comparing the total bond enthalpy of benzene with the value calculated for cyclohexatriene

This is a book-keeping exercise to see how much energy is required to tear apart two gaseous molecules, one having fixed and one delocalised bond structures. Remember that bond enthalpies refer to the energies required to break internal bonds, not intermolecular bonds. That is why we must start with benzene in the gaseous state. We can calculate the total bond enthalpy by two routes. The process is

$$C_6H_6(g) \rightarrow 6C(g) + 6H(g)$$

Route 1 We start with gaseous benzene and convert it into atoms indirectly.

(i) Convert the gaseous benzene into liquid benzene in its standard state:

$$C_6H_6(g) \rightarrow C_6H_6(l) \qquad\qquad \Delta H_1 = -31 \text{ kJ mol}^{-1}$$

(ii)   then create the elements (in their standard states):

$$C_6H_6(l) \rightarrow 6C(s, graphite) + 3H_2(g) \qquad \Delta H_2 = -49 \text{ kJ mol}^{-1}$$

This is the reverse of the formation of benzene where $\Delta H_f = +49 \text{ kJ mol}^{-1}$. (Unusually, the formation of benzene is endothermic.)

(iii)   then atomise the elements:

$$6C(s, graphite) + 3H_2(g) \rightarrow 6C(g) + 6H(g) \quad \Delta H_3 = +5598 \text{ kJ mol}^{-1}$$

Here, $\Delta H_3 = 6(+715) + 3(+436)$

The sum of these processes is what we want to achieve, so that:

$$\Delta H = +5598 - 49 - 31 = +5518 \text{ kJ mol}^{-1}$$

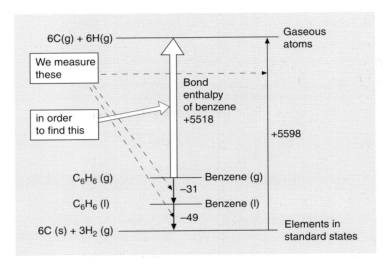

Fig. 3.7 Enthalpy changes / kJ mol⁻¹ leading to the total bond enthalpy.

We have made no assumption about the structure of benzene; we know that 5518 kJ is required to disrupt 1 mol of gaseous benzene into atoms whatever its structure might be.

Route 2 Let us calculate the bond enthalpy of cyclohexatriene using the average bond enthalpies of its fixed bond structure:

Total bond energy	=	$3E(C=C) + 3E(C-C) + 6E(C-H)$
	=	$(3 \times 611) + (3 \times 346) + (6 \times 413)$
	=	$1833 + 1038 + 2478$
	=	$+ 5349 \text{ kJ mol}^{-1}$

The known figure for benzene, as we have seen in Route 1, is $+5518 \text{ kJ mol}^{-1}$, so we need $(5518 - 5349) = 169$ kJ more energy to break up 1 mol of gaseous benzene than we would to break up 1 mol of gaseous cyclohexatriene.

The figures for the stabilising effect of delocalisation, or the 'resonance energy of benzene', by the two routes are 152 and 169 kJ mol^{-1}. The agreement is good when you consider the large totals which the numbers involved would give. The discrepancy reflects:

(i) the use of mean bond enthalpy terms to calculate a total bond enthalpy. This is probably a safe use because we are not looking at the disruption of an isolated bond in a structure. Errors are likely to be very small.

(ii) the uncertainty of the enthalpy (change) of atomisation of carbon, about which there is some disagreement. An error of only 3 kJ mol^{-1} in 715 (less than half of one percent) would, when multiplied by 6 [Route 1, part (iii)], account for all the difference.

Reason for substitution reactions, rather than addition reactions

The one remaining problem associated with the structure of benzene can now be resolved. The benzene ring is thermodynamically very stable as we have shown in the previous section and this stability is associated with the delocalisation of the pi molecular orbital. Addition reactions to benzene would result in disruption of this delocalisation and so reduce the stability of the benzene ring. Substitution reactions can occur with only temporary disruption and so the stability of the benzene ring is maintained.

Aromatic compounds

As we have seen, benzene appears from its formula to be very unsaturated. It might be expected to show alkene addition reactions with great readiness, but it does not. Only when 'forced' will benzene show addition reactions, e.g.

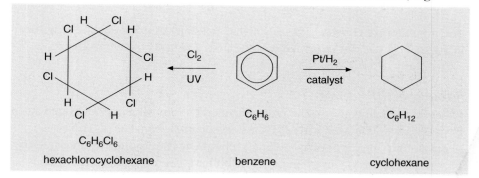

A wide range of other compounds and their derivatives show these characteristics, e.g. naphthalene and pyridine (Figure 3.8).

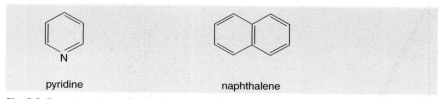

Fig. 3.8 Parent compounds of other aromatic series.

BENZENE – THE BASIS OF AROMATIC COMPOUNDS

Such compounds are described as **aromatic**. Other compounds, by contrast, are described as **aliphatic**. The only aromatic compounds in the Edexcel syllabus are benzene derivatives.

Naming compounds of benzene

Most compounds of benzene are named by indicating the substituent in front of benzene, e.g.

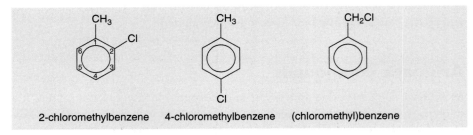

If there is only one substituent it does not matter to which carbon atom you attach it. If the substituent is connected directly to the benzene ring by a carbon atom, then the substituent is known as a **side chain** and the ring as the **nucleus** or, more loosely, the **ring**.

If there is more than one nuclear substituent, then the carbon atoms in the ring are numbered consecutively (starting from a side chain) 1 to 6:

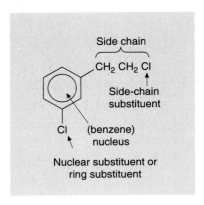

Fig. 3.9 Parts of a benzene derivative.

2-chloromethylbenzene   4-chloromethylbenzene   (chloromethyl)benzene

Fig. 3.10 Three of the isomers of C_7H_7Cl.

The numbers are chosen to be as small as possible, thus the second structure in Figure 3.11 has the same name as the first, it is not 1,6-dichlorobenzene.

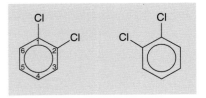

Fig. 3.11 The same compound, 1,2–dichlorobenzene.

Some special names

Phenyl is the name give to the radical formed by removing one hydrogen atom from benzene. It is used with –NH_2 instead of aminobenzene. $C_6H_5NH_2$ is phenylamine (in older books 'aniline'). Phenyl, $C_6H_5–$, is often abbreviated to Ph; do not do this on an exam paper if you have been asked to write the structure of an aromatic compound.

You would expect the radical from benzene to be called benzyl; unfortunately this older name is given to the phenylmethyl- radical, $C_6H_5CH_2-$.

Toluene

Toluene is the old name given to methylbenzene. These old names had some advantages. If someone spoke of chlorotoluene, you knew that the chloro was a ring substituent, whereas if they spoke of benzyl chloride you knew that the chlorine was in the side chain of the same skeleton. If someone speaks of chloromethylbenzene today you can never be sure that they have omitted a number intentionally or just forgotten to say it (see Figure 3.10).

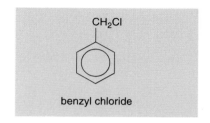

benzyl chloride

Alkyl, aryl and acyl

Alkyl we have met in an earlier unit. It is the radical obtained from an alkane, C_nH_{2n+1}, often symbolised by R. Aryl is the general term for any aromatic radical obtained by removing the hydrogen atom from the nucleus, e.g. phenyl and chlorophenyl- would be aryl groups. Aryl is symbolised by Ar. Acyl is the group obtained by removing –OH from a carboxylic acid RCO– or ArCO–.

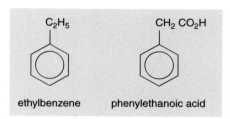

ethylbenzene phenylethanoic acid

Health, safety and aromatic compounds

Many aromatic compounds are associated with health hazards. Benzene is a possible cause of leukaemia and the carcinogens in cigarette tar contain aromatic rings. On the other hand, half the 'links' in the DNA chain are aromatic; every protein and enzyme in your body contains aromatic rings. In fact, the only abundant organic compounds in your body which do not contain aromatic rings are the much maligned fats and sugars!

Benzene is the best choice to learn about simple aromatic chemistry but it should be avoided in laboratory practical work.

Simple substitution reactions of benzene

Nitration

Mixtures of concentrated sulphuric and nitric acids introduce the $-NO_2$ substituent to benzene. The reactions usually have to be cooled initially, but heated to achieve completion:

nitrobenzene

Nitro compounds are widely used to make aryl amines. Several explosives are based on nitro compounds, e.g. TNT.

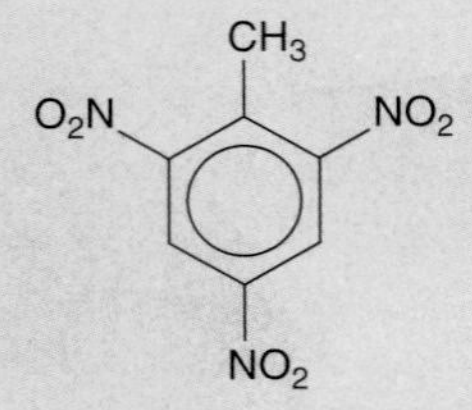

BENZENE – THE BASIS OF AROMATIC COMPOUNDS

TNT

Bromination

Bromine requires a catalyst of iron wire and dry conditions to substitute in the benzene ring. Initially the iron is converted to iron(III) bromide. Heat is not normally necessary. Indeed, the bromine is normally added in small amounts to the aromatic compound and iron wire in a flask, with a condenser fitted for reflux in case the mixture becomes too hot.

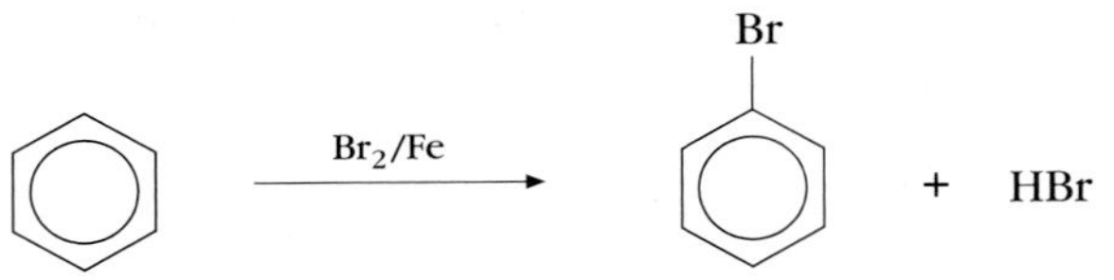

bromobenzene

The resulting aryl halides, unlike haloalkanes (alkyl halides), are unreactive towards sodium hydroxide or silver nitrate. They do, however, form Grignard reagents.

Alkylation and acylation

Both alkyl and acyl groups can be introduced to form a side-chain by the Friedel–Crafts reaction. The aromatic compound (benzene) and the alkyl halide (haloalkane) or acyl halide are mixed in the presence of anhydrous aluminium chloride under anhydrous conditions. The reaction usually works well at room temperature but, like nitration, may need heating (under reflux) to complete:

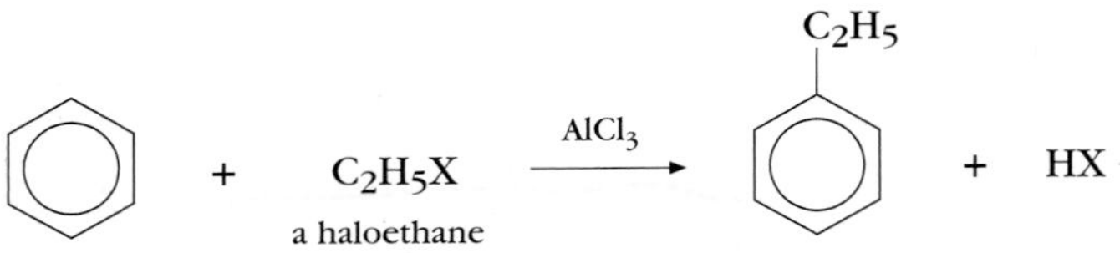

a haloethane

ethylbenzene

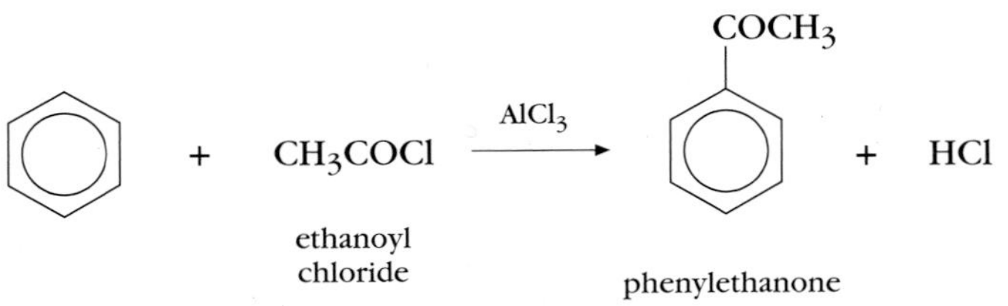

ethanoyl chloride

phenylethanone

The catalyst has to be cautiously destroyed by adding cold water.

Side-chain oxidation

Gentle side-chain oxidation achieves the same results as in the aliphatic compounds. Thus acidified sodium or potassium dichromate will oxidise 2-phenylethanol (**A**), first to phenylethanal (**B**) and then to phenylethanoic acid (**C**).

$$C_6H_5CH_2CH_2OH \quad \rightarrow \quad C_6H_5CH_2CHO \quad \rightarrow \quad C_6H_5CH_2CO_2H$$
$$\mathbf{A} \qquad\qquad\qquad \mathbf{B} \qquad\qquad\qquad \mathbf{C}$$

Note, though, that you should not use Ph– or C_6H_5– if you are asked to draw a full structural formula in an exam paper.

Similarly, phenylpropan-2-ol (**D**) would be oxidised to phenylpropanone (**E**), which, in turn, could be oxidised by the haloform reaction to phenylethanoic acid (**C**) and triiodomethane CHI_3:

$$PhCH_2CH(OH)CH_3 \quad \rightarrow \quad PhCH_2COCH_3 \quad \rightarrow \quad PhCH_2CO_2H$$
$$\mathbf{D} \qquad\qquad\qquad \mathbf{E} \qquad\qquad\qquad \mathbf{C}$$

Hot alkaline potassium manganate(VII) solution oxidises all side chains *in which carbon is directly linked to the benzene ring* to carboxyl groups –COOH. Thus all five of the compounds **A**–**E** above would be oxidised to benzoic acid (benzenecarboxylic acid), $C_6H_5CO_2H$. This reaction is of greater diagnostic than preparative use. A C_9 benzene derivative, on oxidation with hot alkaline manganate(VII), will give a mono-, di- or tri- (etc.) carboxylic acid, which will be easily recognised as such, both by its analysis (formula is $[C_6H_6 + nCO_2]$ where n is the number of carboxylic acid groups: $C_7H_6O_2$ = mono-, $C_8H_6O_4$ = di-, $C_9H_6O_6$ = tricarboxylic acid), and also by its melting point (and mixed melting point) or infra-red spectrum. Correct identification of a benzenedicarboxylic or -tricarboxylic acid will give immediate information on the number and relative positions of the side chains that have been oxidised. Benzene-1,2-dicarboxylic acid may also be recognised because it loses water very easily on heating:

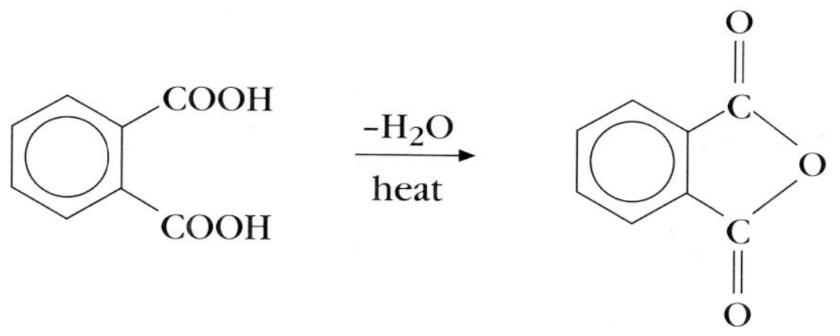

Phenols

These compounds contain the –OH group connected directly to the benzene nucleus. The parent compound is called phenol; others are named as derivatives of phenol.

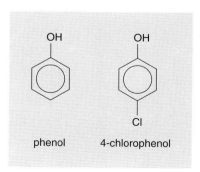

phenol   4-chlorophenol

Although phenol is made industrially by the action of hot aqueous sodium hydroxide on chlorobenzene at high pressure this is not a suitable laboratory method of introducing the hydroxy group to a benzene ring. The most common route is via the nitro compound, e.g.

$$C_6H_6 \quad \rightarrow \quad C_6H_5NO_2 \quad \rightarrow \quad C_6H_5NH_2 \quad \rightarrow \quad C_6H_5OH$$

You will learn about these steps later. The conditions have been omitted now to allow you to test yourself when you revise.

BENZENE – THE BASIS OF AROMATIC COMPOUNDS

The –OH group in phenols is more acidic than in alcohols. Although phenols are not very soluble in water they are soluble in aqueous alkalis, e.g. sodium hydroxide:

phenol   phenate ion

QUESTION

K_a (phenol) = 10^{-10} mol dm^{-3}. Write an expression for this and use it to calculate the pH of 0.10 mol dm^{-3} phenol.

This is because the anion produced is stabilised by delocalisation of the negative charge. The orbital on the oxygen which is responsible for the negative charge overlaps with the π-system of the benzene ring (compare Figure 3.13).

Aqueous solutions of phenols seldom have a pH below 5. Phenols can be recognised because they are soluble in strong alkalis and are re-precipitated by strong acids but are insufficiently acidic to decompose sodium carbonate or, in most cases, sodium hydrogencarbonate. Unlike carboxylic acids, they do not liberate carbon dioxide.

Phenol itself is a solid at room temperature with a strong 'antiseptic' smell. In contact with water its melting temperature is so depressed that it usually forms an oil. Experiments with phenols must be done wearing gloves and eye protection. Industrially, anyone handling it or its aqueous solutions in bulk must wear waterproof clothing. It gives vicious burns to the skin and is highly toxic.

Historically, in the nineteenth century, its use by Joseph Lister revolutionised 'antiseptic surgery'. So successful was it that, at one stage, it was the practice to spray the air in the operating theatre with aqueous phenol to kill any air-borne germs. This irritating mixture was not popular with surgeons of the day.

Phenol shows enhanced activity for substitution in the 2, 4 and 6 positions. A solution of bromine is decolourised by phenol with the formation of a white precipitate of 2,4,6-tribromophenol. A well known test:

2,4,6-tribromophenol

The reaction is both rapid and quantitative so that the concentration of a solution of phenol can be found by titration. A measured amount of aqueous bromine is added to the solution of the phenol and, after a few minutes, potassium iodide solution is added to react with the unused bromine. The liberated iodine is titrated with sodium thiosulphate. Phenols are similarly reactive towards chlorine. Chlorophenols are extensively used as antiseptics

(TCP) and preservers of wood and packing paper. They are the precursors of a range of selective weedkillers.

Phenols are more difficult to esterify than aliphatic alcohols. The usual method is to shake a solution of the phenol in aqueous sodium hydroxide with the appropriate acyl chloride in excess.

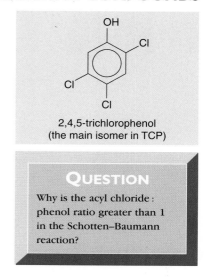

2,4,5-trichlorophenol
(the main isomer in TCP)

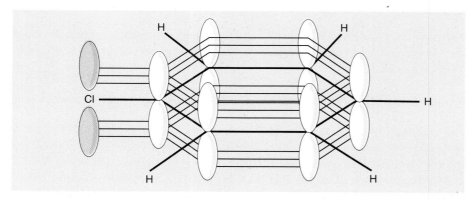

Fig. 3.12 *Esterifying a phenol.*

After the reaction is complete the residual acyl chloride is hydrolysed by the alkali. The procedure is known as the Schotten–Baumann reaction.

QUESTION

Why is the acyl chloride : phenol ratio greater than 1 in the Schotten–Baumann reaction?

Amines

Preparation of aromatic amines

The reduction of aromatic nitro compounds is the usual route to aromatic amines. This route is, in theory, applicable to aliphatic nitro compounds, but these are not so easy to make, and there are many easier routes such as nucleophilic substitution in haloalkanes by ammonia, reduction of cyanides or the Hofmann degradation.

Aromatic nuclear chloro compounds are unreactive to nucleophiles like OH⁻ and ammonia. The interaction of the π-electrons with an unshared pair on the chlorine increases the bond order and strength of the C–Cl bond. Thus simple substitution of chlorobenzene (Figure 3.13) by ammonia is not possible.

Fig. 3.13 *Orbital overlap in chlorobenzene.*

Aromatic nitro compounds can give rise to a wide variety of different reduction products: the method of reduction must be chosen with some care. The traditional laboratory reductant is tin and concentrated hydrochloric acid:

$$C_6H_5NO_2 + 6H^+ + 6e^- \rightarrow C_6H_5NH_2 + 2H_2O$$

with

$$Sn + 6Cl^- \rightarrow SnCl_6^{2-} + 4e^-$$

showing that, to balance the overall equation, 3 mol of tin ($= 12e^-$) are required for 2 mol of nitrobenzene.

Not only is tin a relatively expensive metal (and its large atomic mass means that a lot is required) but also large quantities of alkali are needed to neutralise the large amount of concentrated acid used before the amine is released from the complex hexachlorostannate formed, $C_6H_5NH_3^+HSnCl_6^-$. The method is unsuitable industrially for economic reasons.

The weakly basic aromatic amines form salts with acids, e.g. $C_6H_5NH_3^+Cl^-$. These simple salts, which are a means of selling amines in a convenient and easy-to-handle form, are not suitable for the identification of amines by melting point since they invariably decompose before melting. The most accessible stable covalent derivatives are prepared by treatment with an acyl halide (usually ethanoyl chloride, CH_3COCl, or benzoyl chloride, C_6H_5COCl) or an acid anhydride:

$$C_6H_5NH_2 + CH_3COCl \rightleftharpoons C_6H_5NHCOCH_3 + HCl$$
$$\text{\textit{N}-phenylethanamide}$$

These substituted amides are sometimes used as pharmaceuticals, e.g. paracetamol.

Amines and nitrous acid (HNO$_2$)

A solution of sodium nitrite (sodium nitrate(III)) in dilute acid gives the unstable, pale blue nitrous acid or nitric(III) acid, HNO_2. This reacts differently with aliphatic and aromatic primary amines.

With aliphatic amines, very poor yields of alcohols are obtained:

$$RNH_2 + HNO_2 \rightleftharpoons ROH + N_2 + H_2O$$

The reaction is a simple diagnostic test for primary (aliphatic) amines, although the observation of a colourless gas with negative identification tests is not one to inspire confidence.

For aromatic amines, the reaction can be halted at an intermediate diazonium ion:

$$ArNH_2 + HNO_2 + HCl \rightleftharpoons ArN_2^+Cl^- + 2H_2O$$

In order to make use of this intermediate, it must be prepared at below 5°C and in the presence of excess acid. As it is prepared in aqueous solution, this temperature requirement is particularly easy to satisfy by simply adding ice. The low temperature gives the diazonium ion kinetic stability and the excess acid, among other things, prevents it reacting with unchanged amine. It is used *in situ* as soon as possible: on standing, or if the temperature rises, it decomposes to give a poor yield of a phenol (cf. the aliphatic reaction):

OH

NHCOCH$_3$

paracetamol

$$ArN_2{}^+Cl^- \ + \ H_2O \ \rightleftharpoons \ ArOH \ + \ N_2 \ + \ HCl$$

a phenol

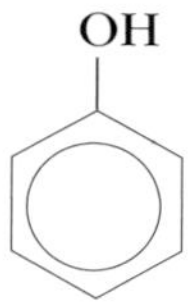

The diazonium ion is an electrophile and condenses with phenols and aromatic amines to give highly coloured compounds. The amines employed are usually tertiary, in order to prevent the electrophile (El) from attacking the lone pair on the nitrogen, and displacing a proton:

OH

phenol

The initial formation of the diazonium ion is called **diazotisation**. The subsequent reaction with the aromatic amine or phenol is called **coupling**:

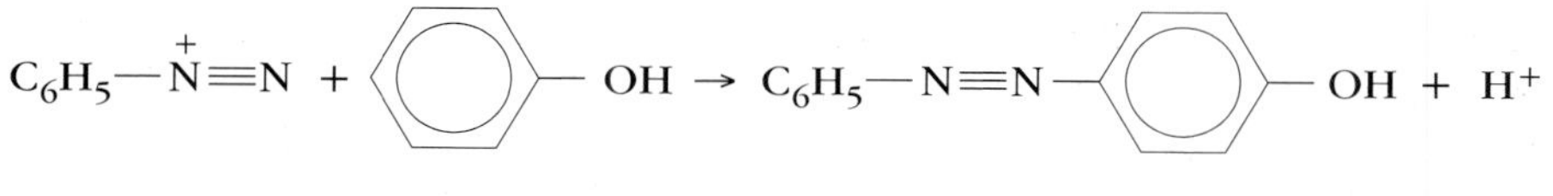

A minor use of the reaction is as a test for aromatic primary amines. An attempt is made to diazotise the suspected primary amine, and an alkaline solution of 2-naphthol is added to the mixture. A positive result is the formation of a highly coloured precipitate, usually red:

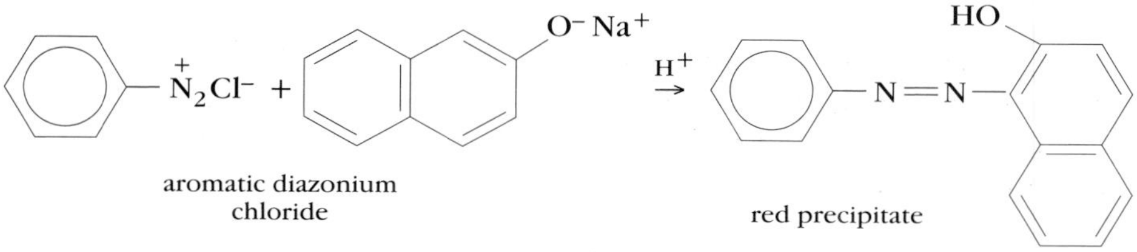

aromatic diazonium
chloride

red precipitate

> ### QUESTION
> Explain why this test cannot be used for the detection of a tertiary amine.

The main use of the reaction is to produce synthetic dyes and some acid–base indicators (see p.105).

Questions

1 How would you convert benzene into (i) bromobenzene, [1] (ii) methylbenzene, [1] (iii) benzoic acid, $C_6H_5CO_2H$, [2] (iv) (chloromethyl)benzene, [2] (v) ethyl benzoate [3]. The numbers in square brackets are suggestions for the number of steps you might use.

2 25.0 cm^3 of a solution of phenol was treated with 25.0 cm^3 of aqueous bromine.

$$C_6H_5OH \ + \ 3\,Br_2 \ \rightarrow \ C_6H_2Br_3OH \ + \ 3\,HBr$$

After five minutes, excess aqueous potassium iodide was added

$$Br_2 \ + \ 2\,I^- \ \rightarrow \ 2\,Br^- \ + \ I_2$$

and the liberated iodine required 5.0 cm^3 of 0.100 mol dm^{-3} sodium thiosulphate solution for reduction

$$I_2 + 2 S_2O_3^{2-} \rightarrow 2 I^- + S_4O_6^{2-}$$

When the same procedure was carried out using water instead of phenol solution, 45.0 cm^3 of the same sodium thiosulphate solution was required. Calculate the concentration of the phenol solution. You may wish to use the following steps:

  (i) What amount of bromine remained after reaction with phenol?

  (ii) What amount of bromine was present before the reaction?

  (iii) What amount of bromine reacted with the phenol?

  (iv) What amount of phenol was used?

  (v) What was the original concentration of phenol?

3 An organic compound A, $C_9H_{11}I$, on treatment with aqueous potassium hydroxide solution gave, among the products, a compound B, $C_9H_{12}O$. B responded to oxidation in three different ways. With acidified potassium dichromate it yielded C, $C_9H_{10}O$. With sodium hydroxide and iodine it yielded D, $C_8H_7O_2Na$ and a yellow precipitate E. With hot alkaline potassium manganate(VII) it yielded F, $C_7H_6O_2$.
(i) Give the structures of A, B, C, D, E and F.
(ii) Which of the compounds, A to F, is capable of showing optical isomerism?

When A was treated with hot ethanolic potassium hydroxide, a mixture of isomers, G, C_9H_{10}, was isolated.
(iii) Draw the structures of the three possible isomers of G. Label the pair of stereoisomers and indicate how their names would distinguish between them.

Reaction mechanisms

Reaction mechanisms tell us: which bonds are broken; how they are broken; which bonds are made, and in what order; and for when reactants form products. They help us understand: the conditions we must use to favour reactions; the types of compound from which a reaction is likely to be successful; and, when a reaction occurs at a chiral centre, whether the stereochemistry will be retained.

Bond breaking

A covalent bond is a shared pair of electrons. This can break in one of two ways:

- **Heterolytic fission:** Both of the shared electrons are given entirely to one of the atoms which shared them. This always results in the creation or transfer of charges:

$$A–B \; \rightarrow \; A^+ \; + \; :B^-$$

- **Homolytic fission:** One electron of the shared pair is given to each of the covalently linked atoms. This results in the creation of at least one free radical:

$$A–B \; \rightarrow \; A\cdot \; + \; B\cdot$$

Heterolytic reaction mechanisms use curved arrows to indicate the movement of pairs of electrons. The arrow should start at the lone pair or at the middle of the bond. If a bond is being formed, the point of the arrow should finish between the bonded atoms; if the bond is being broken and the electrons form a new lone pair, the point of the arrow should finish at the atom or ion with which the lone pair is associated. For example, in a nucleophilic displacement:

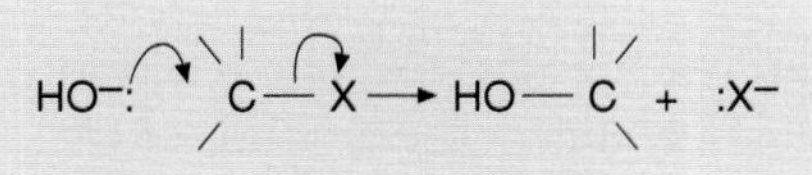

Homolytic reaction mechanisms sometimes use a 'fish hook' to represent the movement of one electron. For example, in the (gaseous) dissociation of iodine

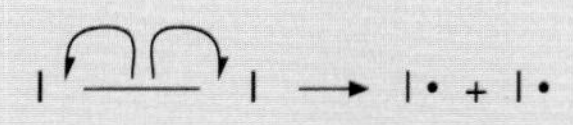

Chlorination of methane

The mechanism for the chlorination of methane is **homolytic free radical substitution**. The mechanism for this reaction is as shown below:

> **DEFINITION**
>
> **Free radicals** are species which have a single unpaired electron

the initiation step:

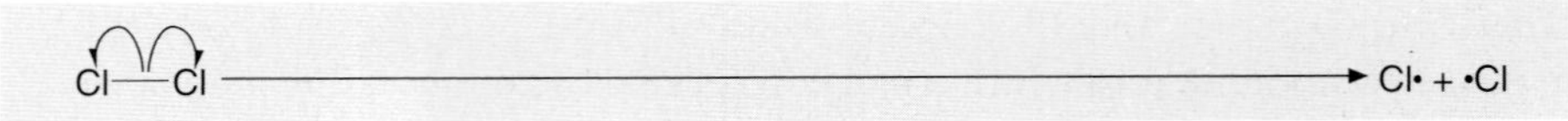

Cl—Cl
Cl• + •Cl

the propagation steps:

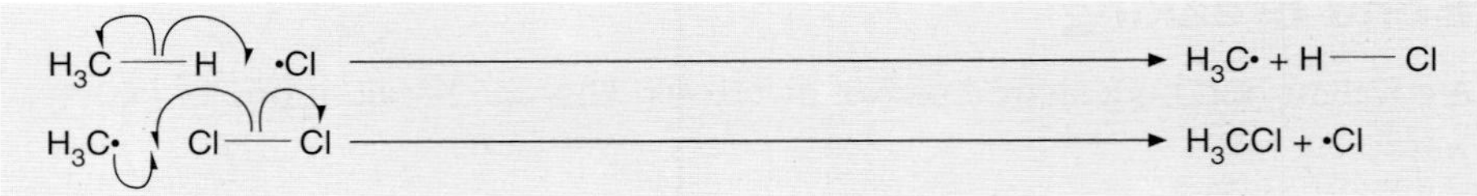

H3C—H •Cl
H3C• + H—Cl
H3C• Cl—Cl
H3CCl + •Cl

possible termination steps:

$$Cl\bullet + \bullet Cl \longrightarrow Cl_2$$

$$Cl\bullet + \bullet CH_3 \longrightarrow ClCH_3$$

$$H_3C\bullet + \bullet CH_3 \longrightarrow H_3CCH_3$$

The **initiation step** is the reaction which first generates the free radicals. A lot of energy is required to break this bond, which is why UV light is necessary.

The **propagation steps** are reactions which are instigated by free radicals which, on reaction, regenerate replacement free radicals, so allowing the reaction to continue unaided.

The **termination steps** are reactions in which free radicals are used up and not regenerated. As a result, the reaction stops eventually.

Most free radicals are extremely unstable and reactive species. The reaction between chlorine and methane occurs explosively unless the amount of UV radiation is controlled by using subdued sunlight.

Even with the correct mole ratio CH_4:Cl_2 =1:1, other products such as dichloromethane are formed by the competing reaction of the initially formed chloromethane with chlorine.

Homolytic polymerisation

The scheme for homolytic polymerisation is shown in Figure 4.1. The initiator is a free radical, $R^\bullet$, e.g. a peroxide. The new free radical formed can then react with another alkene molecule, to form a longer free radical. Eventually, **termination** of free-radical polymerisation occurs by attack on an established chain. Unfortunately, termination of one chain might very well initiate a reaction in part of an established chain. The new radical then grows 'from the middle'. This tends to produce untidy, branched-chain structures that do not pack well. Since melting point and mechanical strength depend on the force between molecules, and this falls off very rapidly with separation, such branched polymers soften and melt easily, and have poor mechanical strength.

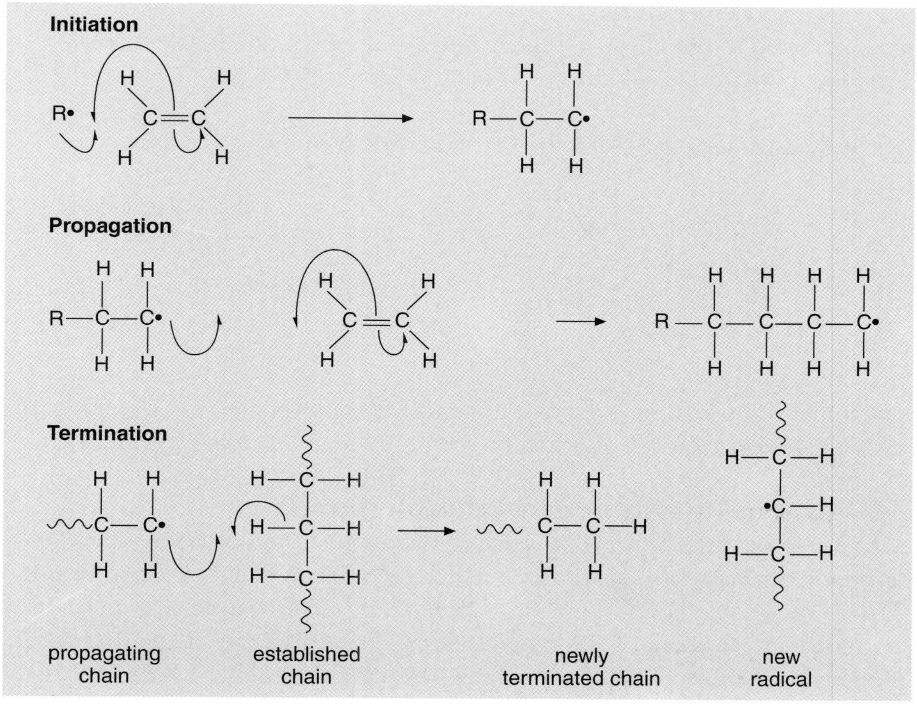

Fig. 4.1 The three stages in homolytic polymerisation.

Addition reactions

Reaction with hydrogen bromide

The reaction with ethene at room temperature is as follows:

$$H_2C = CH_2 + HBr \rightarrow CH_3CH_2Br$$

bromoethane
(a colourless liquid)

Reaction with bromine

Solutions of bromine are decolourised by alkenes. This is a good test for unsaturation. If the bromine is pure or in an organic solvent the typical reaction is:

$$H_2C = CH_2 + Br_2 \rightarrow BrCH_2CH_2Br$$

1,2-dibromoethane
(a colourless liquid)

Both of the reactions given and all the other addition reactions for alkenes occur by a mechanism known as **heterolytic electrophilic addition**. The mechanism for the reaction of ethene with hydrogen bromide is as follows:

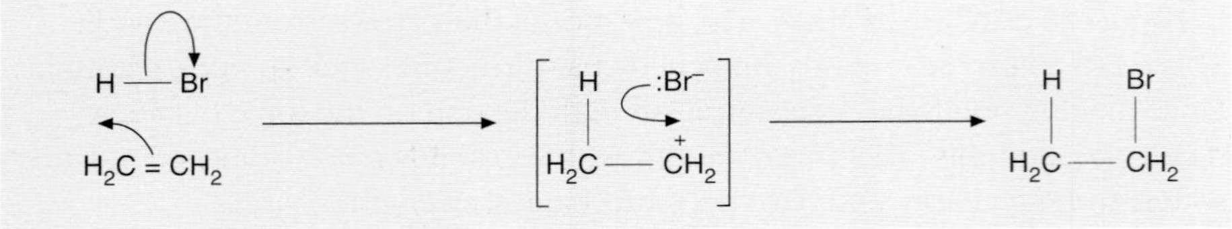

REACTION MECHANISMS

The electrophile here is the $H^{\delta+}$–$Br^{\delta-}$ molecule which is already polarised and so is attracted to the negative region which is the pi bond. In the case of bromine, polarisation is induced on approaching the π–bond.

The mechanism for the reaction of ethene with bromine is as follows:

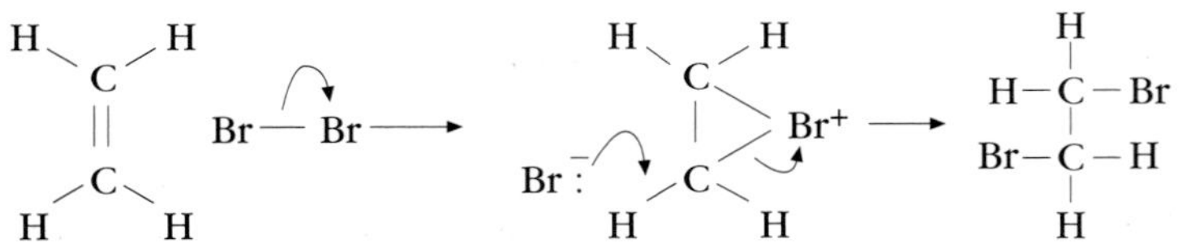

The initial addition of Br^+ forms a triangular π-complex with the π-bond of the alkene. This opens when attacked by the bromide ion forming the product.

Addition at unsymmetrical double bonds

A complication arises on the addition of HX (X = Cl, Br, I or HSO_4) to unsymmetrical alkenes, i.e. those in which the molecule differs on either side of the C=C double bond. It was first observed by Markownikov that, in such additions, the **H atom** of HX joined the carbon atom of the alkene that bore the greater number of hydrogen atoms. This generalisation is known as **Markownikov's rule**. Thus the addition of hydrogen chloride to propene gives predominantly 2-chloropropane:

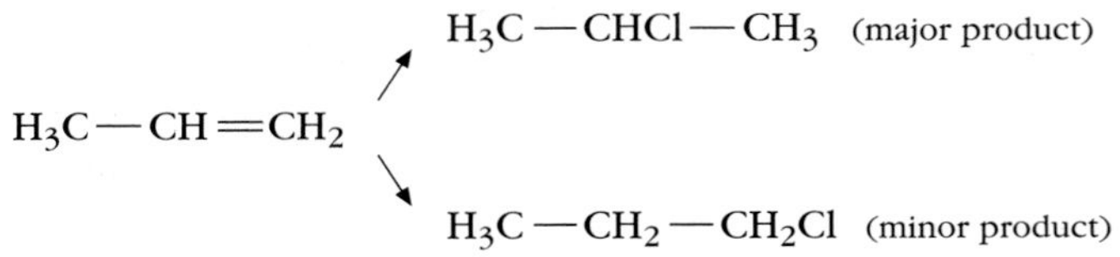

One method of manufacturing propan-2-ol and propanone from propene depends on this preference:

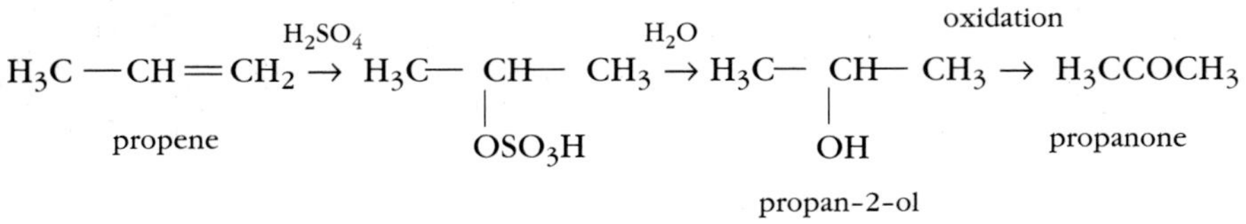

The reason for Markownikov's rule becomes clear if we look at the mechanism. Consider a simple alkene with a terminal double bond. Initially a carbocation is formed by electrophilic attack on the species by HX. Two carbocations are possible:

DEFINITION

A **carbocation** (carbonium ion) is an ion where the positive charge is sited on a carbon atom.

DEFINITION

Markownikov's rule
When a hydrogen halide is added to a C=C double bond, the hydrogen atom is attached to the carbon atom that already carries more hydrogen.

QUESTION

Predict the number of products and name them when HCl is added to pent-2-ene, $CH_3CH{=}CHCH_2CH_3$.

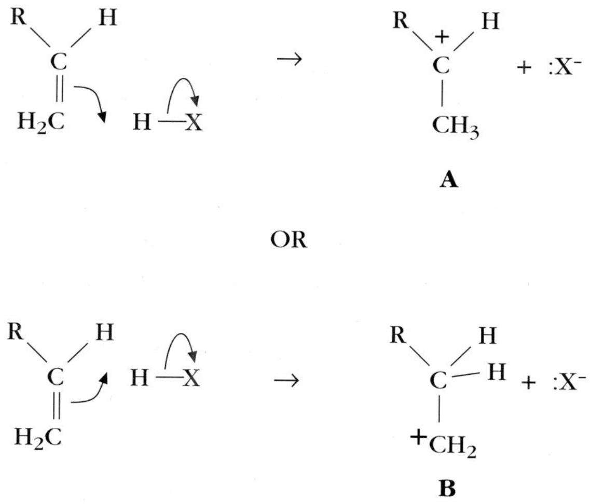

$$R\diagdown C \diagup H,\ H_2C,\ H-X \rightarrow R \diagdown \overset{+}{C} \diagup H + :X^-$$

A

OR

$$R\diagdown C \diagup H,\ H_2C,\ H-X \rightarrow R\diagdown C \diagup H,\ C-H,\ ^+CH_2 + :X^-$$

B

A is a secondary carbocation which is more stable than the primary carbocation **B**. The lower energy of **A** compared with **B** suggests that its formation will require lower activation energy, and thus the formation of **A** will be much faster. The final monochloroalkanes are likely to have similar enthalpies of formation because they have the same bonds present in a different arrangement, but the 2-chloroalkane forms much more quickly and therefore predominates. The reaction is described as **kinetically controlled** rather than **thermodynamically controlled**, as would be the case if the ratio of the products depended on their enthalpies of formation.

For everyday problems it is useful to remember and apply Markownikov's rule. However, it is important to understand that it results from the electrophilic addition mechanism with its carbocation intermediates.

Substitution in the aromatic nucleus
Attack on the nucleus, with its electron-rich π-system, must be by an electrophile. This is classed as (heterolytic) electrophilic substitution. For the Edexcel syllabus, you have to know how each electrophile is generated, as well as how it behaves.

In general, if the electrophile is E^+, initial attack forms a carbocation (carbonium ion) (Figure 4.2).

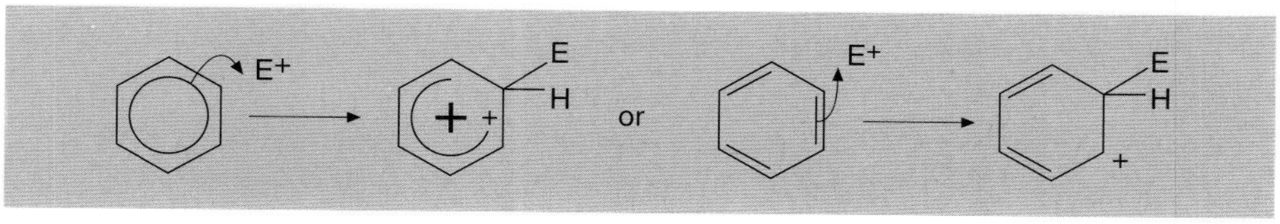

Fig. 4.2 Initial attack of an electrophile.

The electrophile has to be a powerful one because the π-system of the benzene ring is hard to disrupt (see earlier in this chapter). That is why bromine attacks ethene or cyclohexene rapidly but in the absence of a catalyst (or ultraviolet light) has no effect on benzene.

The positively charged intermediate then restores the aromatic π-system by loss of a proton from the same carbon atom attacked by E^+ (Figure 4.3).

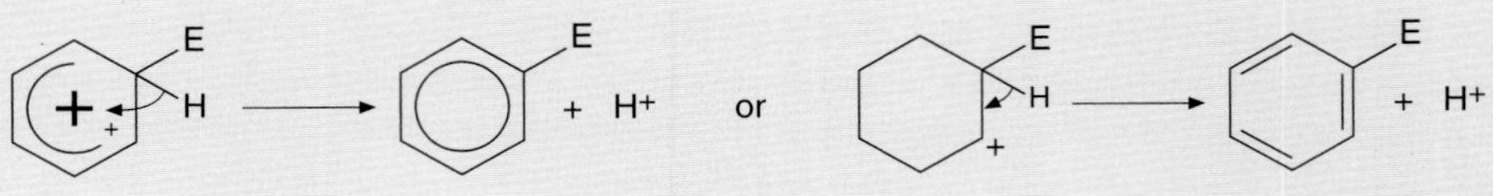

Fig. 4.3 Loss of a proton.

To create an isolated proton requires a lot of energy and the proton doesn't just 'float away'. It requires a base to remove it, e.g. HSO_4^- in nitration, but we normally take that for granted.

The loss of H^+ from the carbonium ion (carbocation) intermediate is energetically more favourable than the addition of an anion, which we observe in alkene addition, because the energy of the aromatic product is much reduced when the aromatic π-system is restored. In alkenes, where the double bond is localised, energy is liberated when it is broken and new single bonds are created.

The Edexcel syllabus allows the use of both the Kekulé structure and the delocalised structure for benzene to explain mechanisms. Many organic chemists prefer to use the Kekulé system because it helps to predict the behaviour of second substitutions. For example, why does further nitration of nitrobenzene give 1,3-dinitrobenzene but very little 1,2- or 1,4-dinitrobenzene? This and similar interesting problems can be explained using the Kekulé structure; they are outside the syllabus. Purists would argue that the structure should never be drawn with fixed bonds, but the structures of naphthalene and anthracene are always drawn this way despite their extended π-systems. Two of the following mechanisms (which are simply repeats of the one above with a specific electrophile replacing E^+) are drawn with Kekulé structures to allow you to decide which you prefer.

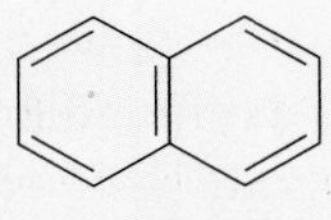

naphthalene 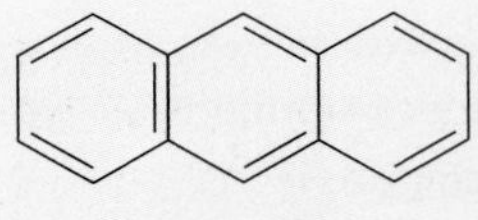   anthracene

Nitration

The electrophile NO_2^+, the nitronium (or nitryl) ion, is generated by the protonation and dehydration of nitric acid (which acts as a base!!!) by sulphuric acid:

$$2H_2SO_4 + HNO_3 \rightarrow H_3O^+ + NO_2^+ + 2HSO_4^-$$

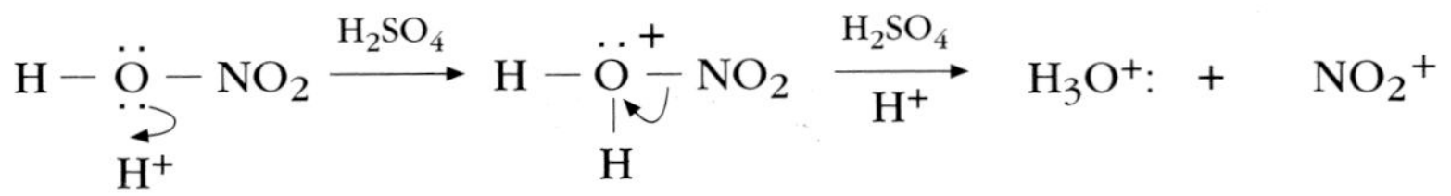

The mechanism is then:

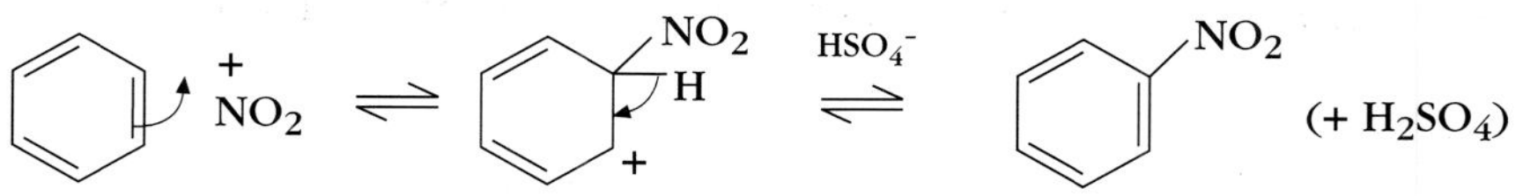

Bromination

The electrophile is Br^+, probably loosely associated with the $FeBr_4^-$ anion. The iron filings first react under the anhydrous conditions to give iron(III) bromide:

$$2Fe + 3Br_2 \rightarrow 2FeBr_3$$

You may, of course, start with an aluminium trihalide instead of iron. The halide forms a complex with bromine

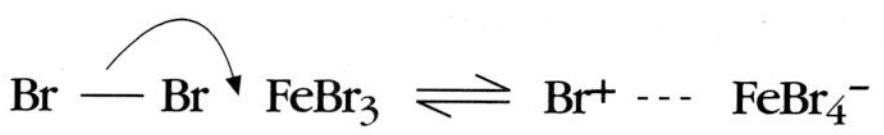

Remember, both aluminium and iron(III) halides are electron deficient when anhydrous (hence the formation of M_2Cl_6 dimers by both). The electrophile will be represented simply by Br^+. The mechanism is:

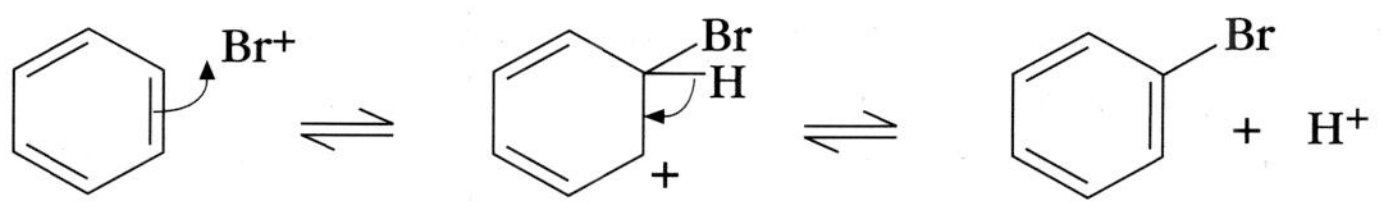

In this case, the H^+ will probably be removed by $AlBr_4^-$

$$H^+ + AlBr_4^- \rightleftharpoons HBr + AlBr_3$$

Friedel–Crafts reaction (alkylation and acylation) (see p. 44)

The electrophile, RCH_2^+ or RCO^+ with the charge on the **bold** carbon atom, is generated in a similar way to Br^+ because the catalysts are essentially the same for both reactions. The electrophile can be represented as the simple RCH_2^+, as we have done with the electrophile Br^+, or you can represent it as the loosely associated ion-pair, $RCH_2^+.....[Cl\text{-}AlCl_3]^-$, as in the mechanism below.

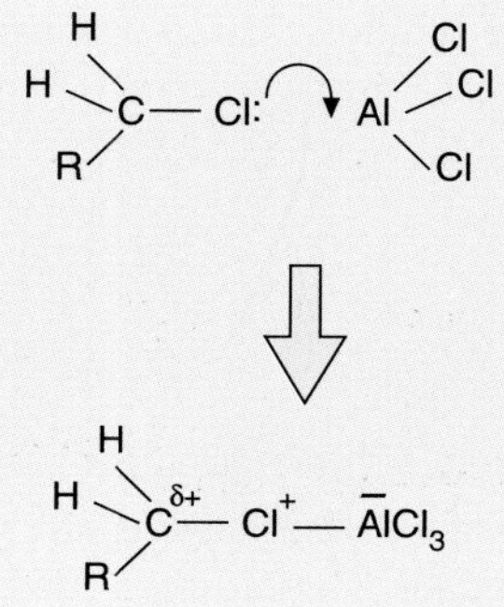

REACTION MECHANISMS

The positive charge generated on the chlorine atom has the effect of pulling the electrons away from the carbon in the C–Cl bond, which makes this C atom an effective electrophile (or, alternatively, makes it very susceptible to nucleophilic attack by the π-electrons of the benzene ring). The complex reacts with the benzene ring and introduces a side chain:

The mechanism with acyl halides is the same; CH_2 in the above is replaced by C=O.

QUESTION

Name $C_6H_5CH(CH_3)CHO$.

A problem with the Friedel–Crafts addition occurs because of isomerisation of alkyl groups. The carbocation intermediate, RCH_2^+, as shown, is a primary carbocation (see later on p. 59). If R is such that a more stable secondary carbocation can be formed, this will probably happen during the addition. Thus 1-chloropropane will substitute in benzene to give a high proportion of $PhCH(CH_3)_2$ which would be the product if you had started with 2-chloropropane.

Heterolytic nucleophilic substitution

The carbon–halogen bond differs from the carbon–carbon and carbon–hydrogen bonds present in molecules of haloalkanes in that it is polarised to a much greater extent because of the greater difference in electronegativity between the halogen atom and carbon atom. The carbon atom therefore has a partial positive charge and the halogen a partial negative charge.

The degree of polarisation will be greatest for chlorine and least for iodine since chlorine has the greater electronegativity. However, this does not determine the reactivity of the halides, which was explained earlier.

$$-\overset{|}{\underset{|}{C}}{}^{\delta+}\!\!-\!\!X^{\delta-}$$

It is this positive charge on the C atom that allows attack by nucleophiles. Both OH^- and CN^- have lone pairs of electrons to donate and therefore can act as nucleophiles. This attack can take place in two different ways.

S_N1 mechanism

S_N1 means that the reaction occurs by substitution of a nucleophile and that it follows first-order kinetics; more precisely, it is unimolecular. This mechanism occurs in two stages.

- Heterolytic fission of the C–Br bond to form a carbocation:

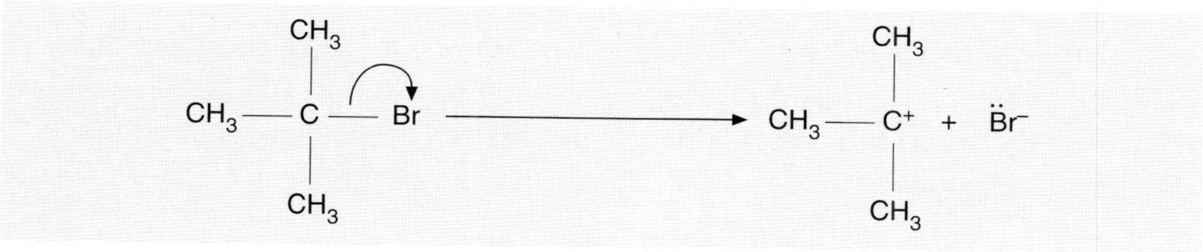

- The nucleophile forms a covalent bond with the carbocation:

The first step of this mechanism is the rate-determining step; the second step is much faster. Hence a study of the kinetics of the reaction would be expected to show that the reaction is first order with respect to the haloalkane and zero order with respect to OH^-. The rate equation would then be:

rate $= k$ [haloalkane]

where k is the rate constant (see Chapter 5). If experimentation confirms this then the mechanism is likely to be S_N1.

Although no attempt has been made to show the shapes of the molecules or intermediates, it should be apparent that the initial and final molecules are tetrahedral but that the intermediate carbocation is planar. This means that the intermediate carbocation can be attacked, by the nucleophile, with equal facility from either side. If the initial haloalkane molecule was chiral, for example 2-bromobutane, equal quantities of the two chiral molecules would be formed and the product would show no optical activity.

S_N2 mechanism

S_N2 means that the reaction again occurs by substitution of a nucleophile but that it follows second-order kinetics. Again, more precisely, it is bimolecular. There is no intermediate in this mechanism. The nucleophile attacks the $C^{\delta+}$ and begins to form a covalent bond with it. At the same time the C–halogen bond begins to break heterolytically. This is one continuous process with the C–nucleophile bond getting stronger and the C–halogen bond getting weaker. The transition state (which corresponds to the activation energy peak) is formed when both bonds are of equal strength.

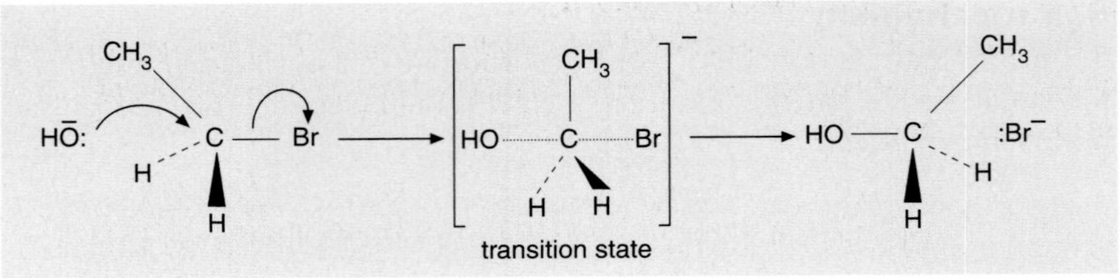

transition state

In this mechanism, the formation of the transition state represents the rate-determining step. A study of the kinetics of the reaction would be expected to show that the reaction is first order with respect to the haloalkane and first order with respect to OH^-. The rate equation would then be:

$$\text{rate} = k\,[OH^-]\,[C_2H_5Br]$$

Will it be S_N1 or S_N2?

Types of carbocation
There are three types of carbocation: primary, secondary and tertiary:

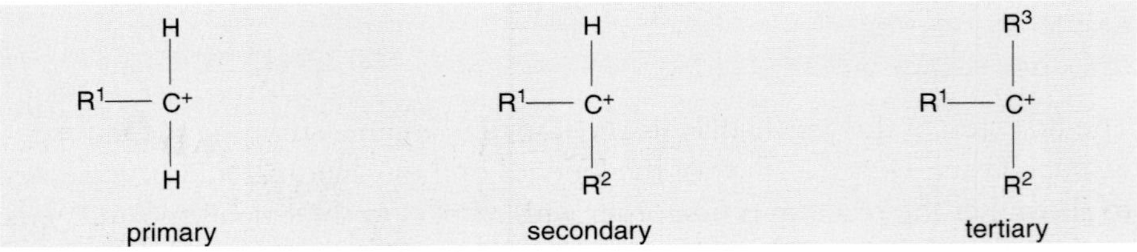

primary   secondary   tertiary

R^1, R^2 and R^3 are alkyl groups which may be the same or different. The terminology is exactly the same as that seen in the various types of haloalkane and will be seen again in other compounds.

The stability of carbocations
Carbocations are unstable species. They can be made more stable if the charge on the cation is dispersed. Alkyl groups are more electron releasing than hydrogen atoms and so push the shared electrons towards the C^+ atom, reducing its charge. The more alkyl groups there are, the more the charge will be dispersed and the greater the stability of the carbocation. Hence the order of stability of carbocations is: tertiary more stable than secondary, secondary more stable than primary. A route involving tertiary carbocations is easier than one involving secondary or primary carbocations.

The only way to be certain which mechanism is occurring in a given reaction is to determine the orders with respect to each reagent experimentally. However, the fact that a carbocation is formed in S_N1 reactions means that tertiary haloalkanes are likely to react via an S_N1 mechanism and primary haloalkanes are likely to react via an S_N2 mechanism. If both mechanisms are operating, as is possible for secondary haloalkanes, the order with respect to OH^- will be fractional.

Nucleophilic reactions at the carbonyl group
Both aldehydes and ketones undergo a wide range of addition and addition–elimination (condensation) reactions.

Nucleophilic addition of cyanide

Hydrogen cyanide is covalent and a weak acid. To provide the cyanide ion, CN^-, an aqueous ethanolic solution of sodium or potassium cyanide containing hydrogen cyanide is used:

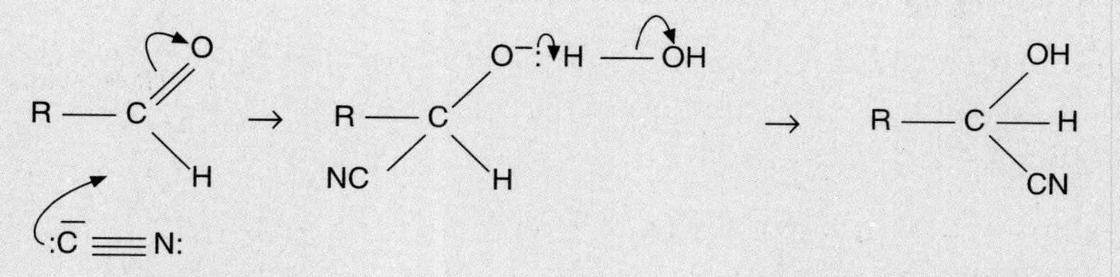

an aldehyde cyanohydrin

Aqueous hydrogen cyanide alone is not a suitable reagent since it is a weak acid.

The resulting cyanohydrins are of synthetic value, as acid hydrolysis gives α-hydroxyacids:

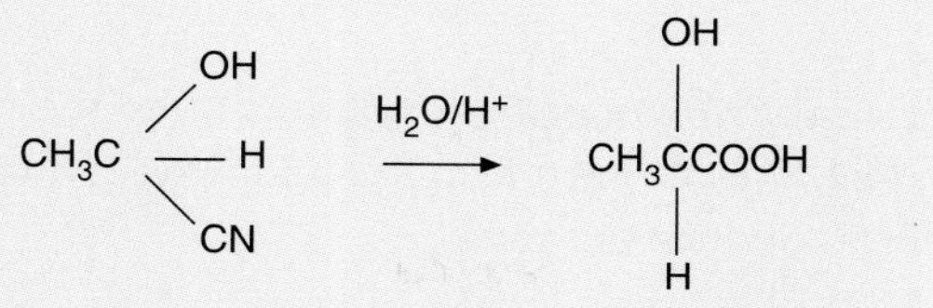

2-hydroxypropanoic acid
(lactic acid)

Questions

1   Explain why the cyanide ion adds to carbonyl groups but not to alkenes, and bromine does the opposite.

2   (a)   Explain why hydrated aluminium chloride, $AlCl_3.6H_2O$, cannot be used as the catalyst in the Friedel–Crafts reaction.

   (b)   How would you use this reaction to produce, from benzene and any other organic materials:

      (i) phenylethanone;

      (ii) 'benzophenone', $C_6H_5COC_6H_5$;

      (iii) 2-phenylpropane ((1´-methylethyl)benzene), $C_6H_5CH(CH_3)_2$?

   (c)   Given benzene and iodomethane as the only organic starting materials, how would you produce methyl benzoate (in several steps)?

3   An unbranched alkene, X, C_5H_{10}, gave two addition products, A and B, with hydrogen bromide. With hot aqueous potassium hydroxide, A gave C and B gave D, both $C_5H_{12}O$. Both C and D were oxidised by excess acidified sodium dichromate(VI) to give E and F, both $C_5H_{10}O$. C and E gave the same yellow precipitate G on warming with sodium hydroxide and iodine; D and F did not react. Identify X and the products A to G and explain your reasoning. Which of these compounds could show stereoisomerism and of what kind?

Kinetics – how fast do reactions go?

Rates, orders and rate constants

For purely mathematical reasons, the rate of a reaction is always expressed as the rate of *decrease* in concentration of one of the *reactants*. If we have a reaction such as:

$$A + 2B + ... \rightarrow 2D + E + ...$$

then the rate of reaction with respect to A, at a time t is given by the expression

$$Rate = \frac{-d[A]}{dt} \approx \frac{-small \ change \ in \ [A]}{small \ time \ taken}$$

From this it follows that the units of rate are $(mol \ dm^{-3})/s$, i.e. $mol \ dm^{-3} \ s^{-1}$.

We can often use the second approximate expression for single measurements of rate.

Of course, the rate of *disappearance* of a reactant is often the rate of *appearance* of a product. It is always a simple multiple (e.g. twice or half) of it. We can measure either to deduce the rate.

Experimentally, we find that:

$$Rate = k \ [A]^m[B]^n$$

where:
- m is the order of reaction with respect to A
- n is the order of reaction with respect to B
- (m + n +...) is the overall order of reaction
- the mathematical relationship is called the rate expression or kinetic expression for the reaction
- m and n might be 1 and 2, as in the reaction stoichiometry, but such a situation is largely chance.

There may be more terms, but this is not common.

Consider these two examples:

Example 1

$$CH_3CO_2CH_3(aq) + OH^- (aq) \rightarrow CH_3CO_2^- (aq) + CH_3OH(aq)$$

The rate of the alkaline hydrolysis of methyl ethanoate is governed by the expression:

$$Rate = k \ [CH_3CO_2CH_3] \ [OH^-]$$

The order with respect to both reactants is 1 and the order matches the stoichiometry.

QUESTION

The concentration of a compound X falls from 0.80 $mol \ dm^{-3}$ to 0.72 $mol \ dm^{-3}$ in 1 min 20 s. Calculate the approximate rate. Why is it almost certainly approximate?

Example 2

Now consider the substitution propanone by iodine in aqueous acid:

$$CH_3COCH_3(aq) + I_2(aq) \rightarrow CH_3COCH_2I(aq) + HI (aq)$$

$$Rate = k [CH_3COCH_3] [H^+]$$

The reaction is first order with respect to propanone, but iodine does not appear in the rate expression – the order with respect to iodine is zero. Instead, the reaction is first order with respect to hydrogen ions which do not even appear in the equation.

The order with respect to any reactant is usually 0, 1 or occasionally 2. The overall order of a reaction is usually 1, 2 or 3. These quantities are *empirical* (found by experiment). Nothing we have met so far in this chapter can be predicted safely from the chemical equation for a reaction alone.

Units of the rate constant

Just as the units of the equilibrium constant, K_c, depend on the equilibrium expression, so the units of the rate constant depend on the rate expression.

For an overall first order reaction:

$$Rate \ (mol \ dm^{-3} \ s^{-1}) = k[A] \ (mol \ dm^{-3})$$

$$units \ of \ k = \frac{(mol \ dm^{-3} \ s^{-1})}{mol \ dm^{-3}} = s^{-1}$$

For an overall second order reaction:

$$Rate \ (mol \ dm^{-3} \ s^{-1}) = k \ [A]^2 \ (mol \ dm^{-3})^2$$

$$units \ of \ k = \frac{(mol \ dm^{-3} \ s^{-1})}{mol^2 \ dm^{-6}} = mol^{-1} \ dm^3 s^{-1}$$

Orders of reaction from initial rate measurements

We do the same reaction, several times, varying only the initial concentration of one chosen reactant. Each time we measure the reaction rate at the start. We can find the order of reaction with respect to the chosen reactant. Suppose we find the rate at the start of a reaction, then we repeat the reaction doubling the concentration of one reactant. If the reaction rate at the start is doubled, ($\times 2^1$), then the order of reaction with respect to our chosen reactant is one. If the rate of the reaction increased by a factor of 4, ($\times 2^2$), then the order of reaction would be 2 and if the rate of reaction was the same, ($\times 2^0$), then the order of reaction (with respect to our chosen reactant) would be zero. We can often use our approximate expression for rate (referring to small changes) in such determinations.

The progress of the next gaseous reaction can be followed by observing pressure changes:

$$2H_2(g) + 2NO(g) \rightarrow 2H_2O(g) + N_2(g)$$

> ## QUESTION
>
> When the initial concentration of X (see previous question) was 0.40 mol dm^{-3} the rate fell to 5×10^{-4} mol dm^{-3}s^{-1}. What is the order with respect to X?

KINETICS – HOW FAST DO REACTIONS GO?

The results of an experiment at 1073 K are given in Table 5.1.

Table 5.1

Experiment	Initial concentration/mol dm^{-3}		Initial rate of N_2 production/ mol dm^{-3} s^{-1}
	[NO]	[H$_2$]	
1	6.00×10^{-3}	1.00×10^{-3}	3.19×10^{-3}
2	6.00×10^{-3}	2.00×10^{-3}	6.36×10^{-3}
3	6.00×10^{-3}	3.00×10^{-3}	9.56×10^{-3}
4	1.00×10^{-3}	6.00×10^{-3}	0.48×10^{-3}
5	2.00×10^{-3}	6.00×10^{-3}	1.92×10^{-3}
6	3.00×10^{-3}	6.00×10^{-3}	4.30×10^{-3}

Experiments 1 and 2 show that when [H$_2$] is doubled and [NO] is constant, the rate of reaction doubles. Hence rate $\propto$ [H$_2$] and the reaction must be first order with respect to H$_2$. A comparison of Experiments 3 and 1 (Table 5.1) confirms this, since trebling the concentration trebles the rate.

Comparing Experiments 4 and 5 shows that when [NO] is doubled while [H$_2$] is constant, the rate increases by a factor of four. Hence rate $\propto$ [NO]^2 and the reaction must be second order with respect to NO. Comparing Experiments 4 and 6 confirm this, since multiplying the concentration by a factor of 3 increases the rate by a factor of 9, that is 3^2. Hence the reaction is third order overall and the rate expression is:

$$\text{rate} = k[\text{NO}]^2[\text{H}_2]$$

Substituting any set of values will then give the value of the rate constant k. Thus using the figures from Experiment 1:

$$k = \frac{\text{rate}}{[\text{NO}]^2[\text{H}_2]}$$

$$= \frac{3.19 \times 10^{-3}}{(6.00 \times 10^{-3})^2 \, (1.00 \times 10^{-3})}$$

$$= \frac{3.19 \times 10^{-3}}{3.6 \times 10^{-8}} = 8.86 \times 10^4 \, \text{mol}^{-2} \, \text{dm}^6 \, \text{s}^{-1}$$

In those very rare situations where the order is not a whole number then we have to use logs. If the initial rates in two 'runs' are rate$_1$ and rate$_2$, then for order n:

$$\frac{\text{rate}_1}{\text{rate}_2} = \frac{\text{conc}_1{}^n}{\text{conc}_2{}^n} = \left(\frac{\text{conc}_1}{\text{conc}_2}\right)^n$$

$$n \log \left(\frac{\text{conc}_1}{\text{conc}_2}\right) = \log \left(\frac{\text{rate}_1}{\text{rate}_2}\right)$$

Rate constants, activation energy and temperature

Most reactions pass through a transition state (TS); one TS for each step if the reaction has more than one step (see below). The TS often represents a condition where old bonds are half-broken and new ones are half-formed. The simplified TS for the alkaline hydrolysis of iodomethane might be as shown (the partial charges would be stabilised by solvation):

$$CH_3I(aq) + OH^-(aq) \rightarrow CH_3OH(aq) + I^-(aq)$$

The TS would be at the top of the 'hump' in the reaction profile (Figure 5.2).

The bigger the activation energy of a reaction, the slower it will be. The rate, and therefore the rate constant, must increase with temperature. The relationship between these variables was first found by Arrhenius (Figure 5.3).

Multi-stage reactions

Simple reactions take place in one step or stage – they are extremely rare. Most reactions are multi-stage, and one step or stage (the words are interchangeable) controls the rate. This is called the rate-determining step and is one important cause of our inability to work out the rate expression from the stoichiometry. Suppose that:

$$A + B + C \rightarrow D + E$$

and that the reaction proceeds in two steps, one slow, one fast:

$$A + B \rightarrow X \text{ (slow)}$$

$$X + C \rightarrow D + E \text{ (fast)}$$

The second stage can be ignored when we look at the rate because A and B will 'queue up' to react and, when they do, the reaction is effectively finished. They will form D and E almost immediately afterwards; [C] will not control the rate. A and B cannot react without colliding, the likelihood of which will depend on how close they are together (the more concentrated they are, the closer they are), hence [A] and [B] in the kinetic expression. The formation of X (an intermediate, not a transition state), is called the rate-determining step. If we found that the rate was first order with respect to A and B but not C, it would be important evidence for such a multi-step reaction. However, it does not prove that the reaction proceeds by this path. In general, we can use kinetics to support a suggested mechanism or reaction path but we cannot deduce from the observed kinetics a unique mechanism.

Measuring rates of reaction

Reaction rates are enormously variable. Some, like the erosion of a chalk cliff, are too slow to measure in the time you have for A-level. Others, like those in fireworks, are too fast to determine at all – certainly with school laboratory equipment. When we can follow the progress of a reaction it is likely to be by analytical sampling of solutions, colorimetry (or UV spectrophotometry), or measuring changes of gas pressure or the volumes of gas evolved from a solution. Whatever method we use, if the results are to have any value, we must

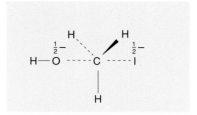

Fig. 5.1 A transition state (simplified).

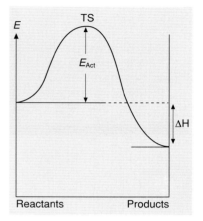

Fig. 5.2 Transition state and forward activation energy.

$$k = Ae^{-\frac{E_{act}}{RT}}$$

Fig. 5.3 The Arrhenius equation.

keep the temperature constant. When we process our measurements, we usually:

- convert them into corresponding concentrations of a reactant
- plot the concentrations on a graph against time
- find the gradient

Sometimes, the order of these operations is varied. The last is required only if we want to know the rate of the reaction for its own sake.

First order reactions and half-life

> ### DEFINITION
> The half-life of a reaction is the time taken for the concentration of a reactant to fall from any selected value to half of that value.

In first-order reactions, if the concentration of the reactant which determines the rate is plotted against time, then the concentration takes the same interval of time to fall to half its concentration, whatever value we start with (Figure 5.4). This enables a first-order reaction to be recognised.

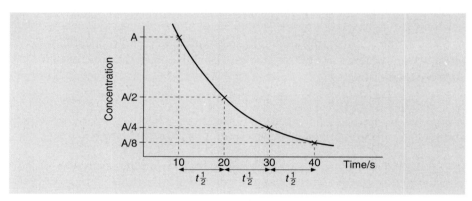

Fig. 5.4 Successive half-lives for a first-order reaction.

The number of techniques for the continuous study of reaction rates is enormous. Three examples are given:

Alkaline hydrolysis of methyl ethanoate

$$CH_3CO_2CH_3(aq) + OH^-(aq) \rightarrow CH_3CO_2^-(aq) + CH_3OH(aq)$$

Equal volumes (100 cm^3) of 0.050 mol dm^{-3} methyl ethanoate solution (freshly prepared) and 0.050 mol dm^{-3} sodium hydroxide are mixed, and timing is begun. The mixture is placed in a constant temperature bath (water at room temperature). At frequent intervals, starting after about 5 min, a 20.0 cm^3 sample is withdrawn. The reaction has to be stopped immediately. If you titrate immediately, the reaction will continue while you titrate. If you add excess hydrochloric acid, acid hydrolysis will commence (while you titrate the acid). If the sample is added to 10.0 cm^3 of ice-cold 0.050 mol dm^{-3} hydrochloric acid, not only will this remove all the remaining sodium hydroxide but also excess hydrochloric acid will be replaced by ethanoic acid, a weak acid which will not hydrolyse the ester, especially at the reduced temperature.

$$CH_3CO_2^-(aq) + H^+(aq) \rightarrow CH_3CO_2H(aq)$$
$$\text{excess}$$

The moment that the hydrochloric acid is added the elapsed time, t, is recorded and the mixture is then titrated with 0.050 mol dm^{-3} sodium hydroxide using phenolphthalein as indicator. The titre, v cm^3, is recorded and the process is repeated.

The concentration of either the sodium hydroxide or the methyl ethanoate at time t is $0.0025(10 - v)$ mol dm^{-3}. Finally plot the concentration of sodium hydroxide (or ester), $[A]_t$, against time (x-axis) (Figure 5.5).

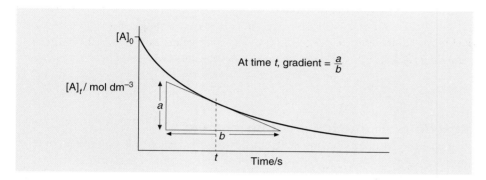

Fig. 5.5 Plot of $[A]_t$ against time. The slope of the curve at any time, t, gives the rate at that time.

The absolute value of rate at any time can be found from the tangent at any point on the curve, as it has the magnitude of the gradient but opposite sign.

You can find the overall order of reaction from your graph but this mathematical treatment is terribly clumsy. The concentration of A (ester or NaOH) is the y-value at the point of the tangent. You can show that the overall order is 2 by drawing two tangents. The ratio of the two rates should be the same as the ratio of the square of the concentrations. The order with respect to either reactant is not necessarily 1. The Edexcel syllabus requires you to find orders from initial rates and, in the case of first-order reactions, from half-life constancy only.

Reactions producing gases

Reactions producing gases are most conveniently followed by measuring the volume of gas produced. Even though the gas is a product and not a reactant, its concentration (or volume) will increase at a rate proportional to that at which the concentrations of the reactants decrease. For example, hydrogen peroxide decomposes in the presence of a catalyst, manganese(IV) oxide, according to the equation:

$$2H_2O_2(aq) \rightarrow 2H_2O(l) + O_2(g)$$

A measured sample of the hydrogen peroxide solution, of known concentration, is placed in a conical flask in a thermostatically controlled water bath, and connected to a gas syringe as shown in Figure 5.6. At a known time, the catalyst

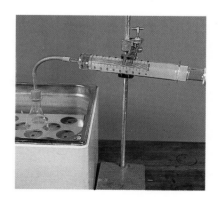

Fig. 5.6 Syringe apparatus for measuring rate of reaction when one of the products is gaseous (see diagram in Figure 5.7).

is added and the volume of oxygen gas in the syringe noted at various time intervals. These volumes are plotted as a function of time, giving a graph as shown in Figure 5.8. The initial rate of the reaction is then proportional to the gradient of this graph at zero time.

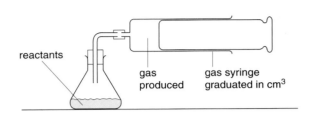

Fig. 5.7 A typical apparatus for measuring volumes of gases produced in a reaction.

Fig. 5.8 Plot of volume of gas evolved against time. The gradient (at zero time) of 0A is the initial rate.

Reactions producing a colour change

If one of the reactants or one of the products of a reaction is coloured, the intensity of the colour can be used to measure the rate of reaction. For example, propanone (CH_3COCH_3) and iodine react together in the presence of an acid catalyst, to give iodopropanone and hydrogen iodide. The equation for the reaction is:

$$CH_3COCH_3(aq) + I_2(aq) \rightarrow CH_2ICOCH_3(aq) + HI(aq)$$

Iodine (as a brown solution) is the only coloured substance in this reaction. Hence the rate of decrease of the colour of the solution is proportional to the rate of decrease of the concentration of iodine and can be used to measure the rate of reaction. The colour intensity can be measured by an instrument known as a colorimeter which records the colour intensity as a reading on a meter or chart recorder. Alternatively, the colorimeter could be interfaced to a datalogger or microcomputer which would allow a graph of concentration of iodine against time to be plotted, as shown in Figure 5.9. This method is particularly useful for fast reactions. The instrument will require calibration in order to establish the relationship between the reading on the meter and the concentration of the species being observed. The rate of this particular reaction

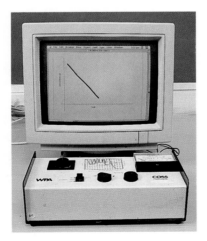

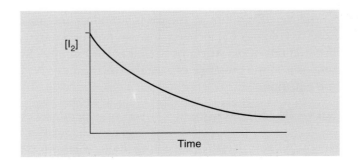

Fig. 5.9 (a) Colorimeter interfaced with computer producing graph of colour intensity against time; (b) results from the datalogger.

could also be followed by the sampling and titration technique described above. The rate of this reaction is such that sampling at 5 min intervals is appropriate. The samples are run rapidly into a flask containing sodium hydrogencarbonate which neutralises the acid catalyst and hence 'quenches' or 'freezes' the reaction. The iodine concentration can then be found at leisure by titration with standard sodium thiosulphate solution using starch as indicator. The colorimeter approach is more satisfactory since no sampling is necessary and the reading can be taken almost instantaneously if a datalogger or microcomputer is used.

Questions

1   The diagram shows the reaction profile for a given reaction.

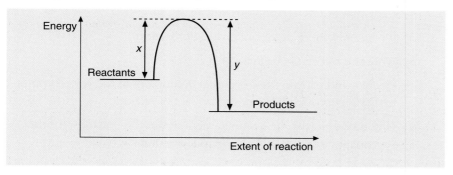

(a) State the following in terms of x and y:

(i) the activation energy of the forward reaction;

(ii) the enthalpy change, ΔH, of the forward reaction.

(b) Make a copy of the diagram and sketch on it a possible energy profile for the same reaction carried out in the presence of a catalyst.

(c) State the enthalpy change, ΔH_1, of the *catalysed* forward reaction in terms of x and y.

(d) Explain, with reference to the diagram, how the catalyst affects the rates of both the forward and the back reactions.

2   The graph below represents the distribution of molecular energies in a gas at a constant temperature T.

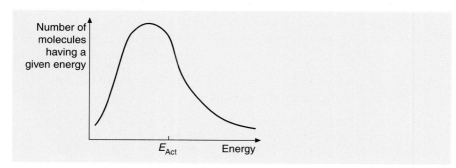

(a) Make a copy of the graph and sketch on it two curves, labelling them T_1 and T_2, showing the distribution of energies at temperature T_1

which is appreciably lower than T, and at temperature T_2 which is appreciably higher than T.

(b) E_{Act} represents the activation energy of a chemical reaction at temperature T. Explain what is meant by the term 'activation energy'. Would its position change for the two curves which you have added to the graph? Explain your answer.

(c) Show how these graphs may be used to explain the effect of an increase in temperature on the rate of a chemical reaction.

3   The stoichiometric equation for the hydrolysis of thioethanamide in alkaline solution is:

$$CH_3CSNH_2 + 2OH^- \rightarrow CH_3CO_2^- + HS^- + NH_3$$

The rate of this reaction is found to be first order with respect to each of the reagents CH_3CSNH_2 and OH^-.

(a) What is meant by the term 'order' of reaction?

(b) Write the rate expression for this reaction.

(c) Deduce and explain the effect that a doubling of the concentration of hydroxide ions would have on the rate of reaction.

(d) Give a reason why the numbers of moles in the equation for the reaction are not the same as the orders for each reagent.

4   In acid solution, bromate(V) ions slowly oxidise bromide ions to bromine:

$$BrO_3^- + 5Br^- + 6H^+ \rightarrow 3Br_2 + 3H_2O$$

The following experimental data was obtained on investigating the rate of this reaction at constant temperature. The solutions of bromate(V), bromide and hydrochloric acid used were all $1 \, mol \, dm^{-3}$.

Mixture	Volume of BrO_3^-/cm^3	Volume of Br^-/cm^3	Volume of HCl/cm^3	Volume of water/cm^3	Relative rate of formation of Br_2
A	50	250	300	400	1
B	50	250	600	100	4
C	100	250	600	50	8
D	50	125	600	225	2

(a) Explain why a certain volume of water is added in each experiment.

(b) Deduce the order of reaction with respect to each of the ions BrO_3^-, Br^- and H^+. Explain your reasoning.

(c) Write the rate expression for the reaction and use it to find the units of the rate constant.

Organic analysis and structure determination

The methods used by organic chemists to find the identity of a simple compound or to find the structure of an unknown compound in the first half of the twentieth century differed little from those in the nineteenth century. The second half of the twentieth century, however, saw a complete revolution in technique.

Before the spectroscopic revolution

Before the advent of spectroscopy, the compound had to be isolated in a pure condition, still an essential requirement when examining a new compound, but techniques of purification have been greatly improved. Solids were separated by solvent extraction and fractional crystallisation; liquids were purified by some form of distillation. Large quantities, of the order of a gram or more, were needed, and the methods of purification gave large mechanical losses.

The first need was to establish the qualitative composition (C, H, O – and what else?) and then the quantitative composition of the compound. Destructive methods were used, e.g. burning a sample of known mass and absorbing and weighing the water and carbon dioxide produced. Further destruction of the precious sample allowed determination of the percentage of nitrogen or sulphur. In skilled hands, reliable results were given, but experimental errors would lend uncertainty to the empirical formulae of complicated compounds like cholesterol, $C_{27}H_{48}O$ (C = 83.5%, H = 12.4%), which might mistakenly be given a formula like $C_{27}H_{47}O$ (C = 83.3%, H = 12.2%) or $C_{27}H_{46}O$ (C = 83.1%, H = 11.9%).

To convert the empirical formula to a molecular formula, a relative molecular mass had to be determined. The methods used were not destructive, but it was not always easy or possible to recover the compound: their main drawback was lack of accuracy. Errors of 10% are not serious if the problem is to decide whether cholesterol has the formula $C_{27}H_{48}O$ ($M_r = 388$) or $C_{54}H_{96}O_2$ ($M_r = 776$): an experimental result of $M_r = 420$ clearly indicates the former. However, to decide whether a sugar with the empirical formula CH_2O was a pentose $C_5H_{10}O_5$ ($M_r = 150$) or a hexose $C_6H_{12}O_6$ ($M_r = 180$) would prove more difficult if the experimental method gave a value of 165.

A long search for functional groups using bromine water, potassium manganate(VII), Fehling's solution, etc., slowly built up a picture of the molecule at the same time as much of the compound was destroyed! You can imagine the problems that faced the organic chemist earlier in this century when deciding how best to use the small amount of material in his or her possession.

In 1935, Doisy processed 1.5 tonnes of sows' ovaries to obtain 12 mg of oestradiol. In his time, the only hope of determining the structure of such a compound with such an amount was to show that it was identical with one already known (see below).

If a compound did not appear to be identical with anything already known, the compound had to be broken down chemically into recognisable molecular pieces, or degraded, by 'nibbling off' one functional group after another, into something recognisable. All the problems of loss of material and poor yields, of separation and purification, began again.

QUESTION

If the M_r is about 420, why is a molecular formula of $C_{27}H_{47}O$ impossible?

ORGANIC ANALYSIS AND STRUCTURE DETERMINATION

When a molecule or molecular fragment was thought to be familiar, its properties had to be compared with known compounds. The melting point of a solid or the boiling point and refractive index of a liquid had to be found. But millions of organic compounds exist with melting points and boiling points in the practical range of, say, 20 to 300°C, and that means that thousands are occupying every degree of the scale.

Solids were easier to identify than liquids. The probable identity of a solid of known melting point could be confirmed with a fair degree of certainty by doing a 'mixed melting point' with a genuine sample. Thus an aromatic carboxylic acid (X), $C_8H_8O_2$, might be one of the following:

CO$_2$H	CO$_2$H	CO$_2$H	CH$_2$CO$_2$H
A	**B**	**C**	**D**
mp 180°C	111°C	108°C	76°C

QUESTION

What would you use to oxidise **X** and what would you expect to get from the two possible structures?

An observed melting point of 108°C would almost certainly mean that **X** was 2-methylbenzenecarboxylic acid (**C**, mp 108°C), but the presence of a very small amount of impurity in **X** might have lowered its mp, and the possibility that **X** was 3-methylbenzenecarboxylic acid (**B**, mp 111°C) could not be ruled out. (Had the determined mp of **X** been 111°C, however, C would be most unlikely.) Small samples of **X** would be separately mixed with an equal bulk of compounds **B** and **C**, and the melting point of the mixtures would be determined. If the mixture with **B** began to melt at, say, 95°C but the mp of the mixture with **C** was largely unchanged, the unknown compound, **X**, would almost certainly be **C**. You *could* decide between these two possible structures by oxidation of **X** and deter-mination of the mp of the oxidation product, but it would be more trouble and would require more material.

E	**F**
bp 170°C	bp 171°C

G	**H**
mp 94°C	mp 110°C

72

Liquids were usually more difficult to identify than solids. To begin with, accurate boiling points are more difficult to determine and require more material than melting points: mechanical losses tend to be larger when handling liquids. If a liquid molecular fragment (**Y**), $C_7H_{12}O$, was recognised to be a saturated ketone with a boiling point of $170 \pm 1°C$, it might be: 3-methyl*cyclo*hexanone (**E**) or 4-methyl*cyclo*hexanone (**F**). Derivatives were prepared, e.g. phenylhydrazones (**G** and **H**). These solids could be narrowed down to a few possibilities by considering their melting points in addition to the information already known.

If the phenylhydrazone of the unknown compound had a melting point of 94°C, that would indicate that **Y** was 3-methyl*cyclo*hexanone (**E**). To confirm the identity, either the phenylhydrazone of a known sample of **E** would be prepared and a mixed melting point carried out, *or* a second derivative, e.g. the 2,4-dinitrophenylhydrazone, would be prepared.

If **Y** (= **E**) had been obtained by the oxidation of a double bond in a more complicated molecule, e.g. by the use of ozone (trioxygen), and the compound formed simultaneously had been butanone, the chemist then had to find a means of deciding how the two fragments fitted together. Do they form **I** or **J**?

butanone   **E**   **I**

QUESTION

What kind of isomers are I and J?

butanone   **E**   **J**

Sometimes, when structures were broken down chemically, rearrangements of the skeleton occurred and chemists were misled by the isomers that they isolated.

While chemical methods like these are now largely displaced by spectroscopy, *chemical* analysis is still indispensable in certain fields. Unbelievably rapid methods have been devised for the automatic sequencing of bases in DNA, and for the amino acids in proteins, which replace methods that took, literally, years. However, these methods are based on chemical reactions.

ORGANIC ANALYSIS AND STRUCTURE DETERMINATION

Summary of tests

In a school laboratory, the purpose of tests for functional groups is to illustrate simple chemistry and allow you to exercise your judgement. Their use in modern structural determination would be far too wasteful of the substance being investigated. You have met the tests in various units and the tests are summarised below; no theory is given.

Group	Test	Comment
C=C alkenes	bromine solution is decolourised cold alkaline potassium manganate(VII) goes green (or brown)	given by some other reducing agents
C–Cl C–Br C–I	haloalkanes: white or pale yellow precipitates of AgCl, AgBr or AgI with ethanolic silver nitrate acyl halides: RCOX: rapid reaction with water to give acidic solutions which react with silver nitrate as above aryl halides: unreactive	primary are slowest not required by Edexcel
C–OH	alcohols: give steamy fumes of HCl with PCl_5. Also give pleasant smelling esters with carboxylic acid using conc H_2SO_4 as a catalyst 1°, 2° : turn acidified potassium dichromate(VI) green	beware wet samples 1° give aldehydes and acids 2° give ketones ($KMnO_4$ also reduced)
RCOR RCHO	ketones and aldehydes give an orange precipitate with 2,4-dinitrophenylhydrazine solution aldehydes reduce ammoniacal silver (oxide) solutions to give a black precipitate or a silver mirror and Fehlings solution to give a red precipitate of copper(I) oxide	 aldehydes reduce stronger reducing agents, of course, like manganate(VII)
$RCOCH_3$ $RCH(OH)CH_3$	give a yellow precipitate (CHI_3) when treated with sodium hydroxide and iodine or potassium iodide and sodium chlorate(I)	requires careful control of conditions. Iodoform is given by both ethanol and ethanal.
RCOOH	pH<<7: gives carbon dioxide with sodium hydrogencarbonate	

Spectroscopy

A note to the reader. This section occupies many pages. This is not disproportionate to the importance of spectroscopy in chemistry but its size should not be allowed to mislead you as to its importance in the Edexcel specification. A trivial reason for its size is the area of paper required to display spectra. There is a more important reason. For most of the work you do, you can visit a library and find material to back up or extend your study; simple introductions with plenty of examples in spectrometry are not widely available. The examples given in university introductory textbooks are often difficult. This introduction is intended to be self-standing and to bridge the gap between 'A-level' and more advanced introductions. Most future Edexcel questions on this topic are likely merely to involve the matching of IR absorption frequencies to functional groups or of chemical shifts to proton environments (in the context of some structural problem); however, you are more likely to do this with confidence if you have greater understanding. Above all, in a topic that is new to you, you need plenty of examples and practice questions. **You are not expected to remember any absorption frequencies**.

ORGANIC ANALYSIS AND STRUCTURE DETERMINATION

Spectroscopy revolutionised chemistry in the second half of the twentieth century. The main reasons why it is so superior to the classical 'chemical' methods of structure determination, analysis and identification are that:

- results are obtained quickly.

- only minute quantities of materials (milligram quantities) are required.

- methods are often non-destructive.

- results (e.g. M_r) are usually very precise.

The four branches of spectroscopy most useful to the organic chemist are indicated in Figure 6.1, along with their particular ranges of application. The older techniques of ultraviolet (UV) and infra-red (IR) are gradually being superseded for structure determination by those on the right-hand side of the figure, which were technically very difficult to introduce and thus became available comparatively late in the twentieth century.

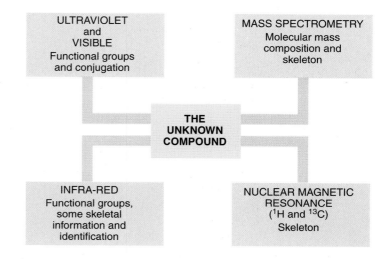

Fig. 6.1 Four branches of spectroscopy that the organic chemist finds most useful.

Absorption spectra

IR spectrometers measure the amount of electromagnetic radiation transmitted by a compound (or a solution of that compound). Two identical beams of IR radiation are produced. One beam is passed through a compound and the other is passed through air, and the emerging beams are compared.

When an **isolated atom** (or ion) of an element absorbs radiation (usually UV), it does so at a precise frequency (or wavelength). As a result, the atom is raised from one electronic state to a more excited state. The energy levels of these two states are precisely defined (E_0 and E_1) and so, therefore, is their difference ΔE (Figure 6.2). This produces the sharp line-spectra of atoms (see *Structure, Bonding and Main Group Chemistry*).

Now while the electronic levels of isolated atoms are precise, those of molecules are grouped about ideal values, because the molecules are in a variety of states of mechanical vibration and distortion, which increase or decrease the total energy. Instead of a precise line in the spectrum of the molecule, a **band** of absorption is produced, centred about the most common energy difference (Figure 6.3). The wavelength of maximum absorption may also be shifted slightly by the environment of the molecule, e.g. the solvent used to dissolve it.

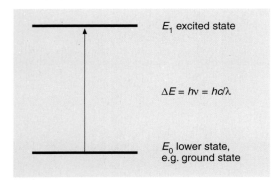

Fig. 6.2 Formation of sharp lines in atomic spectra.

E_1 excited state

$\Delta E = h\nu = hc/\lambda$

E_0 lower state, e.g. ground state

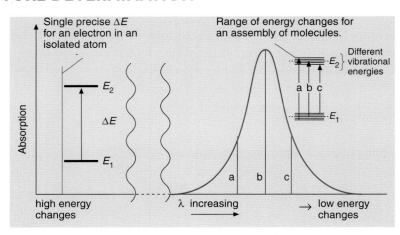

Fig. 6.3 Formation of bands in molecular spectra.

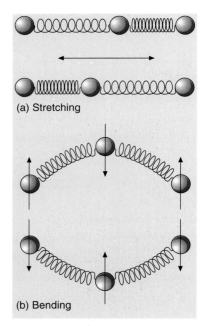

(a) Stretching

(b) Bending

Fig. 6.4 Vibrations of molecules can be thought of in terms of balls and springs.

Infra-red spectroscopy

The energy transitions responsible for infra-red absorption (or emission) are concerned with the mechanical vibrations of molecules, e.g. stretching, or several kinds of bending (Figure 6.4).

The stretching of a C–H bond in an alkane absorbs at a frequency of about $8.7 \times 10^{13}\,\text{s}^{-1}$. Numbers like this are not easy to write or talk about, and spectroscopists have adopted the use of the **wavenumber** to describe the frequency of maximum absorption in the spectrum. If a wave were to be emitted for exactly one second, then the wave train would be $3 \times 10^{10}\,\text{cm}$ long (the speed of light is $3 \times 10^{10}\,\text{cm s}^{-1}$). The wavenumber is the number of wavelengths in $1\,\text{cm}$. In Figure 6.5, drawn for simplicity, the wavenumber would be $5\,\text{cm}^{-1}$ – far outside the IR region, which, for practical purposes, extends from about 4000 to $700\,\text{cm}^{-1}$.

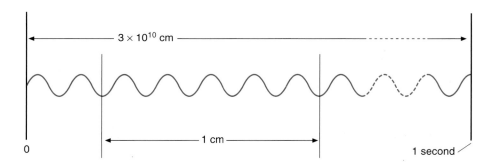

Fig. 6.5 How to think of wavenumbers.

Spectra are usually drawn showing transmittance rather than absorption, and hence the 'peaks' are upside-down. For purely historical reasons, the wavenumber decreases from left to right (Figure 6.6).

Any given bond within a molecule can be stretched or bent in a variety of ways. Hence a given bond can absorb at several different frequencies. Also, as in other vibrations (e.g. in stringed instruments), overtones occur. This makes IR spectra very complicated. Organic chemists are not so much interested in the type of vibration, e.g. stretching or bending, causing a particular absorption, but rather in the ability to associate particular absorption frequencies with *particular bonds in different environments.* Thus a C=C bond in ethene (Figure 6.7) is like a spring with 14 mass units at each end. The 'spring' would have a natural period of vibration (stretching or bending).

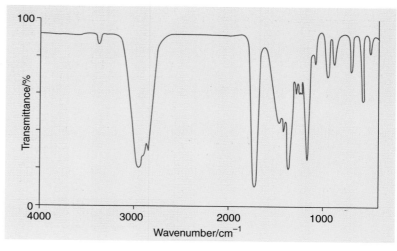

Fig. 6.6 *A typical IR spectrum (that of pentan-2-one).*

The same C=C double bond ('spring') would be expected to have different mechanical properties in oleic acid (*cis*-octadec-9-enoic acid) (**A**), with a mass of over 100 units on each end, or in *cyclo*hexene (**B**), where the double bond or 'spring' is restrained by the rest of the molecule (Figure 6.8). In a cyclic molecule, the rest of the molecule has to distort with every vibration of the 'spring', and the period and energy required are altered.

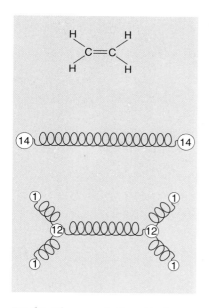

Fig. 6.7 *Ethene can be thought of as a spring with 14 mass units at each end.*

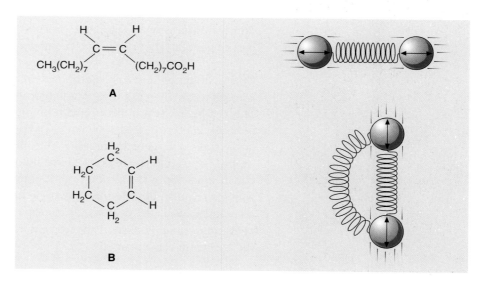

Fig. 6.8 *The same bonds in different molecules will have different mechanical properties and so have different frequencies and spectra.*

Divisions of the IR spectrum

The spectrum can be divided into two working parts, the **band region** and the **fingerprint region** (Figure 6.9).

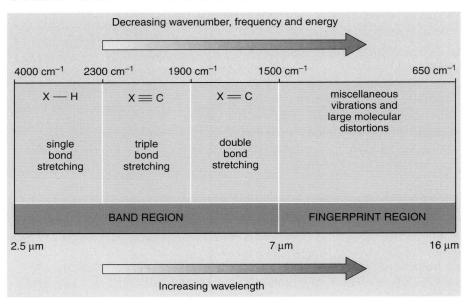

Fig. 6.9 The regions of the IR spectrum.

Plotting the IR spectrum

This is now largely an automatic process. 'IR machines' have been a routine part of general laboratory equipment since the mid-1950s.

Note: a common and easy method of preparing a solid for the IR spectrometer is to grind the material with 'nujol' (commonly also known as 'liquid paraffin'). This is a hydrocarbon mixture with no multiple bonds. It absorbs strongly between 2800 cm^{-1} and 3000 cm^{-1} (C–H) and between 1400 cm^{-1} and 1500 cm^{-1} (skeletal/fingerprint region). This limits the use of such spectra; they cannot be used in the regions where nujol absorbs, so you cannot draw any conclusions about the structure or identity of a compound from this region. Nujol is popular, despite this drawback, because organic compounds generally contain far too many C–H bonds to make the C–H region useful or comprehensible. Published spectra, e.g. those on the Internet, will state how the spectrum was obtained.

Identifying a compound with IR

IR spectra are most valuable for the identification of compounds by comparison of the spectrum of the unknown with those of likely possibilities. Vast libraries of such spectra are maintained by research laboratories, and 'atlases' of these spectra have been marketed in books and in electronic format, e.g. CD ROMs. The spectra are becoming increasingly available on the Internet, sometimes as free downloads. A good source is http//webbook.nist.gov/chemistry for example. This method of identification is far more reliable than the use of melting points, mixed melting points, boiling points and other physical properties.

Identification is not just a matter of comparing the peaks caused by known functional groups, e.g. C=O or C=C. If a known carbonyl compound is being identified, it will be no surprise to find an absorption peak associated with the C=O bond! Rather, the test is to match the absorption in a complex region between 1400 and 1800 cm^{-1} called, for obvious reasons, the fingerprint region. In this region, a huge range of skeletal vibrations produce an absorption pattern of great complexity, which is unique to a given compound.

Investigating structure with IR

The process consists, in essence, of matching the peaks in the **band region** of the spectrum with particular features, e.g. C=O, C=C, O–H, to see which of the possible isomers of a compound is most likely. Shifts in the peak frequencies, sometimes slight, give clues about the environment of the functional group. Even the C–H absorptions of terminal CH_3– and internal –CH_2– are slightly different. Conjugation of double bonds often produces large shifts. Thus the C=O absorption of an aliphatic ketone is normally found at about 1725–1705 cm^{-1} and that of the C=C in a simple alkene is in the region of 1680–1620 cm^{-1}. But in an unsaturated ketone, in which these two π-systems are conjugated to produce an extended π-system (Figure 6.10), the bond orders of the C=C, C=O and intervening C–C cease to be 2, 2 and 1, respectively, and the mechanical properties of the system are altered. The C=O absorption might move to 1685–1665 cm^{-1} and the C=C absorption frequency would similarly be reduced.

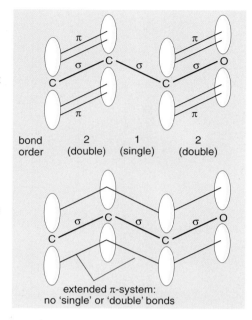

The absorption of the carbonyl group, C=O, also depends (to a lesser extent) on whether it is part of an aliphatic aldehyde (1740–1720 cm^{-1}), a ketone (1730–1710 cm^{-1}), an ester (1750–1730 cm^{-1}), an acid (1725–1700 cm^{-1}) or an amide (1700–1650 cm^{-1}). You are not expected to remember such figures.

Fig. 6.10 Extended π-system in an unsaturated ketone.

The fingerprint region is not of much use in the analytical determination of structure until final identification by matching spectra is reached. However, the absence of an expected band in the fingerprint region can often be diagnostic. For example, the absence of a band at about 1380 cm^{-1} suggests the *absence* of terminal CH_3– groups, but the *presence* of such a band might be due to one of a number of causes and is not a reliable indicator.

Consider three of the many possible isomers of $C_5H_{10}O$, labelled as **A**, **B** and **C** below:

$$H_2C\overset{O}{\diagdown}CH_2$$

A   B: $H_2C=CHCH_2CH_2CH_2OH$ C: $CH_3CH_2\overset{O}{\overset{\|}{C}}CH_2CH_3$

A   **B**   **C**

B should show C=C absorption near 1650 cm^{-1} (although this is often weak) and O–H absorption above 3000 cm^{-1}, whereas **C** should show C=O absorption near 1710 cm^{-1}: **C** should show terminal CH_3– absorption near 1380 cm^{-1}. On the other hand, **A** should show none of these (but, remember, the *presence* of a band in the fingerprint region is not a reliable indicator of the *presence* of a bond).

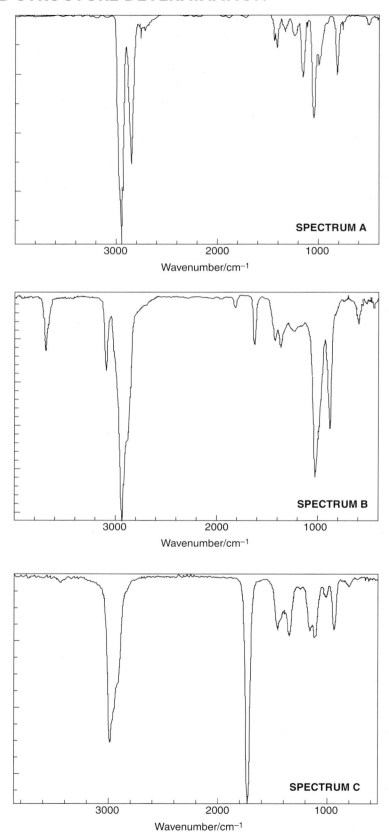

Fig. 6.11 The IR spectra of three isomers of $C_5H_{10}O$.

It would be much more difficult, if not impossible, to distinguish **C** from two other possibilities, **D**, $CH_3CH_2CH_2COCH_3$ (whose IR spectrum is shown in Figure 6.6), or **E**, $(CH_3)_2CHCOCH_3$, without access to predetermined spectra of the pure compounds, but then a simple chemical test would do that. Of course, C=C and C=O, in this case, could have been found by simple chemical tests. A chemist does not use one weapon alone, from the armoury, to fight a battle.

QUESTION

How would you distinguish between **A**, **B**, **C** and **D** by chemical tests alone?

The most common need of IR spectra at A-level is to reveal the nature of oxygen-containing compounds. Is it an aldehyde, ketone, ester, alcohol or acid? C=O stretching occurs at higher frequencies (to the left) for simple aldehydes compared with ketones (Figure 6.12). An expert might look for small differences here, but you would not be expected to do this.

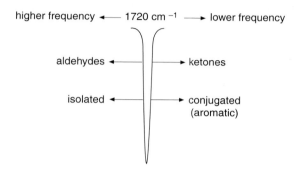

Fig. 6.12 C=O absorption varies.

The difference between aldehydes and ketones is not obvious. C–H absorption is present (obviously) in all organic compounds at about 3000 cm^{-1} but the influence of the C=O on the shared C–H of –CHO causes absorption (two peaks) at about 2800 cm^{-1}. Obviously, you cannot remember these numbers for A-level, but try to think along the lines that C–H in CHO might be different from that in CH_2 and it might be possible to use this to distinguish between the spectra of an aldehyde and a ketone. Compare the spectrum of pentanal (Figure 6.13) with that of a pentanone, e.g. **C**, given earlier.

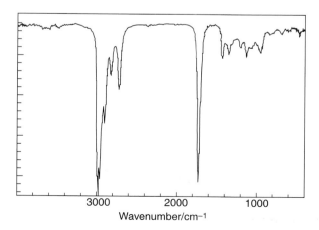

Fig. 6.13 IR spectrum of pentanal.

The O–H absorption is very variable, but is usually distinguishable from the ever-present C–H. The more hydrogen-bonded O–H becomes, the less sharp is the absorption peak and the lower the frequency (to the right). A difference in the O–H frequency of two substances, however, may not reflect a structural difference. It may be that one substance is in liquid form and the other is in vapour form. The presence of an O–H group in a structure can easily be seen, by comparing the IR spectra of methyl propanoate (Figure 6.14) and methyl 2-hydroxypropanoate (lactate). O–H clearly shows in an alcohol or acid (Figure 6.15); the OH peak is sharp and the high frequency because the spectrum is (probably) in the vapour phase.

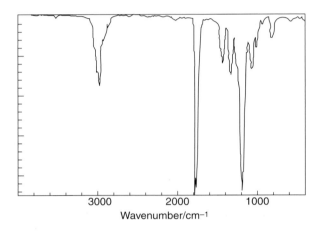

Fig. 6.14 IR spectrum of methyl propanoate.

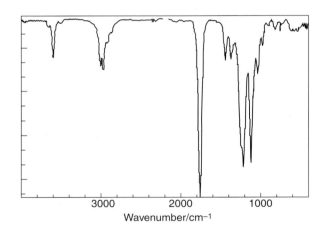

Fig. 6.15 IR spectrum of methyl 2-hydroxypropanoate (lactate).

It is impossible, in an elementary treatment such as this, to give a comprehensive list of absorption frequencies. Some common ones are given in Table 6.1, but you may find rather different values if you look up the ranges in detailed works on spectroscopy.

Table 6.1 *Some common IR absorption frequency ranges*

Bond	Approximate range of frequencies/cm^{-1}	Comments
C–H	3000–2850	in alkanes
C–H	3100–3000	in alkenes and aromatic rings
C–H	900–690	in benzene rings
N–H	3500–3300	primary amines; dependent on hydrogen bonding
O–H	(a) 3600–3300 (b) 3300–2600	broad bands; very dependent on hydrogen bonding, which shifts the absorption to lower frequency (lower energy), i.e. to the right, sometimes beyond the range given extreme H-bonding exhibited by –COOH group broad overlap of CH region.
C–O	1200–1150	in esters
C–Cl	800–600	
C–Br	600–500	
C–I	550–450	
C=C	1680–1620	
C=O	1750–1680	
C≡C	2260–2150	
C≡N	2260–2200	

You will notice that some of these absorption frequencies, e.g. for C–O, are in the fingerprint region. They are unlikely to reveal the presence of the feature because you can never be sure what causes a peak in this region. You are hardly likely to need evidence of the presence of a halogen – if you do, mass spectra (or NMR) are far more helpful. Absorption in this region is sometimes helpful in revealing the presence of a benzene ring because bending of the C–H bond, up-and-down in relation to the plane of the ring, occurs in the region 670–900 cm^{-1}.

Mass spectrometry

A brief description of a (low-resolution) mass spectrometer has been given in *Structure, Bonding and Main Group Chemistry*. The instrument is used in organic chemistry principally for the determination of relative molecular masses, chemical formulae and structures. This form of spectrometry differs from infra-red (IR), ultraviolet (UV) and nuclear magnetic resonance (NMR) in that it does not involve the measurement of absorption of electromagnetic radiation.

Mass spectrometers measure the ratio of mass to charge for positive ions and are calibrated as if they measured mass alone, in atomic mass units. Loosely, therefore, they are said to measure mass, i.e. for convenience, reference to charge is omitted. But it should always be borne in mind that, *in principle*, they measure the mass/charge ratio and, should an ion become doubly charged, a comparatively rare occurrence, it would appear to have half the true mass.

The spectrum is plotted as relative abundance of an ion (y-axis) against mass/charge ratio (x-axis). The x-axis is normally labelled m/e or m/z. In this book we shall use m/z, where z is the number of (electronic) charges (normally 1, which 'disappears'): e has a definite value, which is not 1, and to use true values of m/e would give numbers that are not the same as (or simple sub-multiples of) the relative mass of the particle.

Formation of the spectrum

The common method of generating the positive ions is to vaporise a minute sample of the compound by heating it on a probe in the ionisation chamber (Figure 6.16), which, like all the internal parts of the spectrometer, is under high vacuum. A volatile liquid may simply be allowed to enter this part of the instrument. A heated filament releases electrons, which are drawn across the ionisation chamber to an anode on the other side of the sample vapour. The fast electrons from the filament 'knock out' electrons from the molecules of sample, and the positive ions generated are drawn across the chamber by a weak electric field into a second accelerator chamber through a slit. Once there, the positive ions accelerate rapidly in a strong electric field and emerge through a second slit into a strong uniform magnetic field across a circular cavity, often a 'D' section. Their charge forces them round the curve of the 'D', and only if there is a perfect match between charge/mass ratio and magnetic field will they emerge through the final analyser slit and be detected. The magnetic field (or accelerating voltage) is gradually changed to allow for different masses across the range of the spectrum.

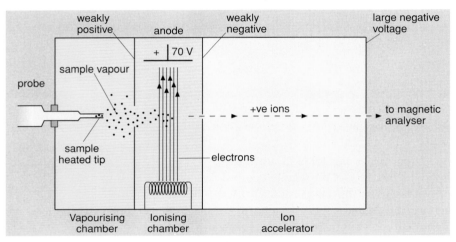

Fig. 6.16 Ion production by electron bombardment in a mass spectrometer.

The nature of the ion

The normal initial reaction of the molecule will be to form a positive ion by loss of one of its electrons:

$$M + e^- \rightarrow M^+ + 2e^-$$

$$\text{high} \qquad\qquad\qquad \text{low}$$
$$\text{energy} \qquad\qquad\qquad \text{energy}$$

The species formed is called the **molecular ion**, and its existence allows the relative molecular mass to be determined. In a high-resolution machine, it may permit the determination of the molecular formula (see below). We shall represent the mass/charge ratio for this ion by M/z and for other ions by m/z.

When the sample is ionised by electron impact, not all the molecular ions survive. Breakdown of the molecular ion in this way is called fragmentation. The instability of the molecular ion arises (like the instability of most chemical species) from two causes:

- The ion has a higher (chemical) energy content than alternative particles into which it can break down (thermodynamic instability).

- The ion has absorbed a large amount of energy from the collision with the ionising electron, which it has retained in the form of vibrations. This vibrational energy may well exceed the necessary activation energy to disrupt the structure of the ion (kinetic instability).

In some cases the molecular ion is undetectable. At the other extreme, many aromatic compounds give very stable molecular ions. Generally, the bigger the molecule, the more likely it is to fragment. Since a knowledge of the molecular mass is crucial to structure determination, much of the progress of the last 20 years has been made by the use of more sophisticated methods of producing the molecular ion than the crude method of electron bombardment.

The nature of fragmentation

Fragmentation does not involve collision between molecules. It is highly unlikely that collision with other molecules would occur at such low pressures and the ions themselves are mutually repulsive.

In simple 'bench' chemistry, if lead(II) nitrate is heated in a test-tube, three molecular species are almost exclusively formed:

$$2Pb(NO_3)_2(s) \rightarrow 2PbO(s) + 4NO_2(g) + O_2(g)$$

At very high temperatures we might get some $NO(g)$ as well. We do not get NO_3, Pb or PbN_2 for sound thermodynamic and chemical reasons.

In a similar way, the decomposition of the molecular ion gives rise to a chemically well defined set of fragments, which themselves may break down further in a predictable way. Experience has allowed chemists to deduce rules for this fragmentation, although the number and nature of the fragments is, to some extent, dependent on the energy of the ionising electrons. We can do no more than hint at these, because we are concerned with building up a picture of the whole molecule from its fragments.

The molecular ion may break down in several ways. The most obvious limitation is that it rarely loses parts from the middle alone.

Only one fragment carries the positive charge: the other fragment may be a radical (split off by homolytic cleavage of a bond) or may be a stable molecule such as H_2, H_2O or $H_2C=CH_2$. The carrier of the charge is not randomly chosen but is predictable, and the neutral fragments, whether radicals or molecules, are not detected.

This means that if a molecular ion of mass 100 gives a positive fragment of mass 60, it may not give the complementary ion of mass 40.

The likelihood of formation of different positive ion fragments depends on their relative stabilities. Like carbocations (carbonium ions) in S_N1 reactions, the fragments increase in stability as the positive charge is placed on a primary, secondary or tertiary carbon atom. Thus when fragmentation of a hydrocarbon chain occurs, the least likely fragment is CH_3^+ and the least likely peak is at

DEFINITION
Molecular ion
This is the ion formed from a molecule by the loss of one electron.

$m/z = 15$. Thus the absence of a peak at $m/z = 15$ by no means indicates the absence of a terminal methyl group: this is in complete contrast to the IR spectrum.

Heteroatoms (not C or H) play an important part in fragmentation, because they have pairs of unshared electrons (on O, Cl, etc.). It is easier to lose an unshared (non-bonding) electron than to lose one from a bonding pair of electrons at a lower energy level in a σ-bond. Cleavage often takes place nearby, partly because of the disturbance of the electrons in the adjacent bonds; e.g. on either side of C=O in a ketone or –CH(OH)– in an alcohol.

If fragmentation occurs very soon after formation of the molecular ion in the ionisation chamber, before the ion passes into the accelerator, then a clean separation of the fragments will occur. A few ions will fragment while in the ion accelerator or the magnetic field. In such cases the positive ion produced will be lost or detected in an inappropriate part of the spectrum. This mistaken identity is not the norm, and the few ions that are detected in the wrong part of the spectrum give rise to a relatively small amount of noise.

Since the *shape* of the peaks is largely due to instrumental defects and 'noise', and the *position* of the maximum and the area of the peak are all-important, the output of the (low-resolution) machine is passed though a computer and printed as a stick-diagram showing intensity (number of ions) against (nearest) mass number. The computer can be programmed to ignore a preset low level of intensity, e.g. 0.5%, thereby eliminating the 'noise'. The most intense peak, called the **base peak**, is given the value of 100 and the rest are scaled accordingly.

Spectrum of iodoethane – a simple example

Let us examine the spectrum of iodoethane (Figure 6.17), ignoring very tiny fragments.

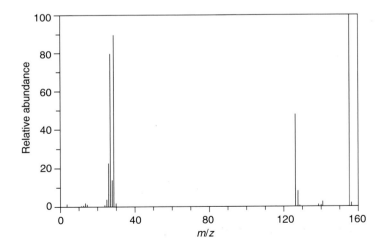

Fig. 6.17 Mass spectrum of iodoethane.

The molecular ion peak $(C_2H_5I)^+$, at $M/z = 156$, is also the base peak. This shows that the molecule is very stable under electron impact conditions. At $m/z = 127$ we have I^+ (not I^-); the peak at $m/z = 128$ is caused by the stable $(HI)^+$ ion. Notice that H is not attached to I in the original molecule but protons have a tendency to move to atoms next to those to which they are attached during fragmentation.

If a molecule loses an iodine atom, there will be no resulting fragment with m/z greater than M–127 (here = 29). A fragment which retains iodine must have m/z greater than 127. The spectrum therefore shows a gap between 29 and 127.

At $m/z = 29$, we have the fragment $(C_2H_5)^+$. As often happens we have a little group of related peaks caused by loss of a hydrogen atom, loss of a hydrogen molecule (very common), or both. Notice how feeble the CH_3^+ peak is.

The next homologue, iodopropane (Figure 6.18a), is clearly not so stable in these conditions but still gives a prominent molecular ion $(C_3H_7I)^+$ at $M/z = 170$ as well as the peaks at 127 and 128 that we saw in the previous spectrum.

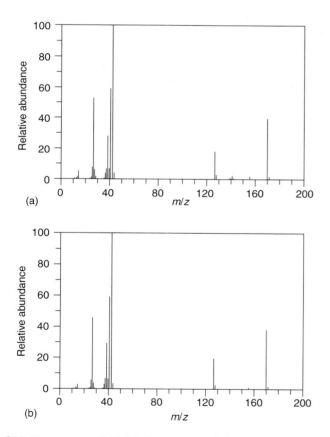

Fig. 6.18 Mass spectra of (a) 1-iodopropane and (b) 2-iodopropane.

The base peak at $m/z = 43$ clearly represents the stable $(C_3H_7)^+$ ion, and the $(C_2H_5)^+$ ion breaks down in much the same way as before. Once again CH_3^+ is not abundant. Notice that we have no strong evidence for 1-iodopropane as opposed to 2-iodopropane. The mass spectrum of 2-iodopropane is so nearly identical with that of iodopropane that, even if transparencies are superposed, the only detectable difference is the absence of the small peak (that we ignored) at $m/z = 141$. It can only be due to $(CH_2I)^+$. This could break away from CH_3CH_2.......CH_2I but not from CH_3CHICH_3. A 4% abundance, on its own, would be rather slim evidence bearing in mind what we have said about protons skipping about to adjacent groups. However, if **both** spectra are present, then on the basis of the **difference**, we have little doubt about the assignment of each spectrum to a structure. We shall see that proton NMR spectrometry, even with only one spectrum, is much more precise than MS.

A note about benzene derivatives

The benzene ring is very stable and the ion responsible for the base peak tends to retain this structure. n-Carbon fragments of an unsubstituted ring should ideally be of mass $13n$ but the devastating effect of smashing the ring causes groups of lines in these regions (Figure 6.19), like the ethyl ion above.

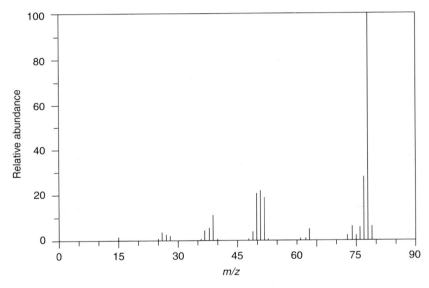

Fig. 6.19 Mass spectrum of benzene.

It is not easy to assign fragments of m/z below 78 to a particular fragment of an aromatic molecule. The region is best ignored if possible.

Isotope peaks

Methanol, CH_3OH, contains three elements, each of which has more than one isotope in the naturally occurring mixture. The most common methanol molecule will be made up of the most abundant isotopes (Table 6.2), carbon-12 (^{12}C), hydrogen (^1H) and oxygen-16 (^{16}O), and will have $M_r = 32$.

However, it is possible to find a molecule, $^{13}C^2H_3{}^{18}O^2H$, $M_r = 39$. The chance of doing this is about 1 in $90 \times (6700)^4 \times 500 = 1$ in 10^{20}: a few thousand molecules in every mole of methanol. This would be lost in the noise of even the most sophisticated mass spectrometer. Only $^{13}CH_3OH$, with a mass of 33, would appear as a tiny peak at $m/z = 33$, commonly referred to as 'M + 1', with an intensity of about 1% of the peak at $M/z = 32$. The molecular ion peak is the one resulting from a molecule made up of the commonest (and usually lightest) isotopes. For a molecule with n carbon atoms, the intensity of the $M + 1$ peak is about n% of the intensity of the molecular ion peak. In elementary questions about mass spectra, it is often the practice to 'clean up' the spectrum by removing such minor peaks.

Peak patterns in the absence of halogens

Peaks, especially the base peak since it is the most intense, usually have weak companions at $m/z + 1$ and even $m/z + 2$ because of the isotope effects described above. This is shown in Figure 6.20. In methane, $CH_4{}^+$, the molecular ion would show a peak at $M/z + 1$ (= 17) of about 1% [theoretically $(1.1 \times 100/98.9) = 1.11$%] of the intensity of the peak at M/z. The molecular ion of ethane, C_2H_6, is twice as likely to have one ^{13}C atom; and the molecular ion for decane, $C_{10}H_{22}$, is ten times as likely, so the peak at $M/z + 1$ for decane should be about 11% of the intensity of that at M/z. On the other hand, the bigger the molecule (unless aromatic), the less likely is the molecular ion to survive, hence the smaller will be the molecular ion peak and its growing isotopic companion.

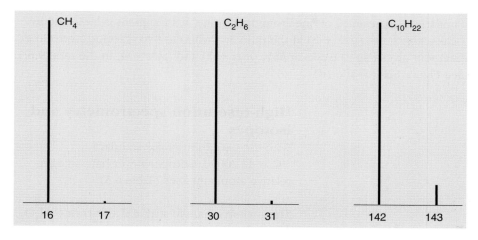

Fig. 6.20 Small peaks at M+1 caused by ^{13}C isotope.

Peak patterns in the presence of halogens

Reference to the table of isotopes (Table 6.2) shows that, while ^{13}C exerts a detectable influence on the spectrum, the presence of chlorine or bromine results in large isotope effects.

If one atom of bromine is present in an ion fragment, then the peak for that fragment will be a doublet of equal intensity separated by 2 mass units. If two atoms of bromine are present in a fragment, it will give rise to a triplet (Figure 6.21). Since the four possible arrangements:

$$[79 + 79] \quad [79 + 81] \quad [81 + 79] \quad [81 + 81]$$

Table 6.2 *Some common isotopes*

Element	Symbol	Mass number	Abundance (%)	Approx. ratio
Carbon	^{12}C	12	98.9	90:1
	^{13}C	13	1.1	
Hydrogen	^{1}H	1	99.985	6700:1
(deuterium)	^{2}H	2	0.015	
Oxygen	^{16}O	16	99.8	500:1
	^{18}O	18	0.2	
Nitrogen	^{14}N	14	99.63	250:1
	^{15}N	15	0.37	
Chlorine	^{35}Cl	35	75.5	3:1
	^{37}Cl	37	24.5	
Bromine	^{79}Br	79	50.5	1:1
	^{81}Br	81	49.5	

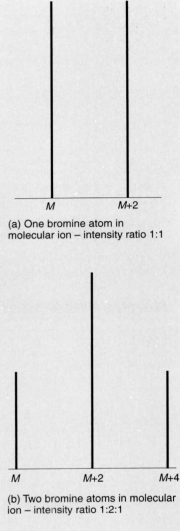

(a) One bromine atom in molecular ion – intensity ratio 1:1

(b) Two bromine atoms in molecular ion – intensity ratio 1:2:1

are equally probable, then the peaks at m/z, $m/z + 2$ and $m/z + 4$ will be in the ratio 1:2:1. It is thus immediately obvious if the molecular ion has two bromine atoms and which fragments have retained both, one or neither of them. See Figure 6.21(a) and (b).

Similarly, the presence of one atom of chlorine in a fragment is indicated by peaks at m/z and $(m/z + 2)$ in the ratio 3:1, whereas two chlorine atoms in a fragment gives rise to peaks at m/z, $(m/z + 2)$ and $(m/z + 4)$ in the ratio 9:6:1. See Figure 6.21(c) and (d).

High-resolution spectrometry and isotopes

With the exception of the standard, $^{12}C = 12.00000$, isotopes never have integer relative atomic masses (Table 6.3).

Thus, while ethanoic acid, CH_3CO_2H or $C_2H_4O_2$, and propanol, $CH_3CH_2CH_2OH$ or C_3H_8O, both appear to have identical M_r values for most purposes, the molecular ions of these species will have masses of 60.0211 and 60.0575. This difference is clearly shown in a high-resolution instrument. A simple computer program will evaluate the possible molecular formulae, given a precise value for the mass of the molecular ion – and all this using quantities of the order of only a milligram.

Building a structure from fragments

This is essentially a jig-saw puzzle with some of the pieces missing. Increasingly, mass

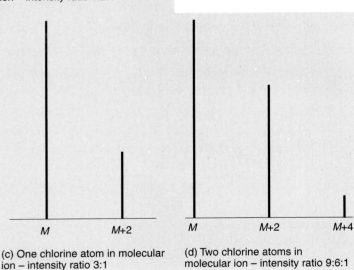

(c) One chlorine atom in molecular ion – intensity ratio 3:1

(d) Two chlorine atoms in molecular ion – intensity ratio 9:6:1

Fig. 6.21 Molecular ion patterns for one and two atoms per molecule of bromine or chlorine.

Element	Mass number of isotope	Accurate atomic mass
Hydrogen	1	1.00783
Carbon	12	12.00000
Nitrogen	14	14.0031
Oxygen	16	15.9949
Sulphur	32	31.9721
Chlorine	35	34.9689
Bromine	79	78.9183

Table 6.3 *Some relative atomic masses*

spectrometry is playing a supporting role to NMR in the determination of complicated structures, rather than leading the research (see p. 95).

Some important points to remember:

- Two fragments of mass greater than $M_r/2$ must overlap.

- Any fragment of mass m is likely to be matched by a fragment of mass $(M_r - m)$ even if this is not present in the spectrum, e.g. terminal methyl groups.

- Peaks differing by 14 mass units usually represent chain length differences of $-CH_2-$, but it is not certain that the less massive (lower m/z) of two adjacent principal peaks has been formed from the more massive one.

- Doublets and triplets are particularly important in establishing the chlorine or bromine content of fragments. Iodine has only one natural species of atom.

- Aromatic fragments (e.g. $m = 77$ for $C_6H_5^+$) are particularly stable. Methyl fragments ($m/z = 15$) are unlikely to be found unless the chain is highly branched, i.e. the number of potential methyl ions compensates for the low probability of their formation.

- Where a very electronegative substituent ($=O$, $-OH$, $-I$, $-Br$, etc.) is present on a carbon chain, fission often occurs on either side of the carbon atom to which it is attached.

- Carbon cannot be more saturated with hydrogen than C_nH_{2n+2}, e.g. C_3H_9 is impossible.

- The molecular ion may *not* be formed for many reasons, but a common oddity of simple spectra is that a hydrogen atom may be lost from an electronegative atom to which it is attached. Thus for simple alcohols the peak at $M - 1$ is often more important than the molecular ion peak. (See the mass spectrum of ethanol in *Structure, Bonding and the Periodic Table*, p. 8.)

Two examples

During our discussion of infra-red spectra (p. 81), we were left with an unsolved problem in which **C**, **D** and **E** could not be distinguished, but must be ketones.

$CH_3CH_2COCH_2CH_3$ $CH_3CH_2CH_2COCH_3$ $(CH_3)_2CHCOCH_3$

C   **D**   **E**

ORGANIC ANALYSIS AND STRUCTURE DETERMINATION

The mass spectra of **C**, **D** and **E** are given below in Figure 6.22 (a)–(c). They are labelled MS1, MS2 and MS3 because you might like to be 'in the dark' as to which compounds they are when you come back to revise this topic.

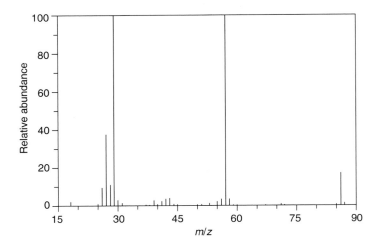

Fig. 6.22a 'MS1'.

MS1 shows the largest fragment at $m/z = 86$ (which must be the molecular ion, because we know the formula). Unusually, there are two base peaks at $m/z = 29$ and 57. The former is clearly caused by $(C_2H_5)^+$ which rules out **E** immediately. (Remember this is a base peak, not an insignificant one.) The rest of **C** has a mass of 57 which matches the other base peak well. Of course, you could get two such peaks by breaking **D** between the two methylene (CH_2) groups, but such an important fission (two base peaks) seems unlikely at an arbitrarily chosen part of a hydrocarbon chain. MS1 is very much what we would expect from **C** if the chain broke on either side of the CO group.

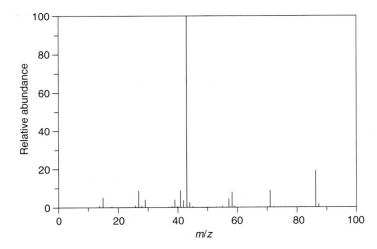

Fig. 6.22b 'MS2'.

MS2 is pleasingly different from MS1. There is one hugely dominating base peak at $m/z = 43$ which seems to be caused by $(CH_3CO)^+$; no doubt, in this case, reinforced by $(C_3H_7)^+$. This rules out compound **C**. Loss of CH_3 to give

$m/z = 71$ could happen to any of the species. A spectroscopist would recognise the loss of C_2H_4 (leaving $m/z = 58$) as characteristic of **D**, but the reason is outside the syllabus; we shall have to ignore it. The peak at $m/z = 29$ (together with its little family) is familiar; there is no $CH_3CH_2–$ in **E**, so we seem to have tied the spectrum rather insecurely to **D**.

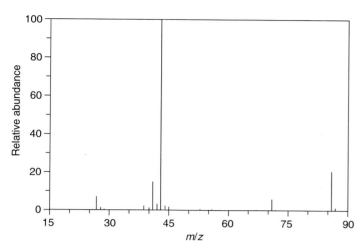

Fig. 6.22c 'MS3'.

Most of the peaks in MS3 are similar to those in MS2. The peak that was at 58 (which we cannot use) is not there. The arrangement of the family of peaks at 29 is different; 29 is particularly weak and 27 relatively strong. We have seen that the ethyl ion tends to lose molecular hydrogen but there is the possibility that we are dealing with the fragment $(CH_3C)^+$ from **E**. If we had all three spectra, we could use the *comparative* evidence to assign them to **C**, **D** and **E** in that order. If we only had MS3, we would be in some doubt. We shall use compounds **D** and **E** again to test the efficiency of NMR spectrometry (see p. 100).

A compound Y of C, H, O and Cl

The formula of **Y** may be known to be $C_{13}H_9OCl$ ($M_r = 216.0342$) because high-resolution mass spectrometry gave a value of 216.03 or it may have been analysed by more conventional combustion analysis.

The main peaks, after removal of the ^{13}C isotope peaks and minor peaks caused by loss of H• or H_2, are listed in Table 6.4.

The molecular ion region (Figure 6.23) is instructive since it shows that (i) $M_r = 216$ and (ii) the molecule contains only one chlorine atom, since the peaks at $m/z = 216$ and 218 are in the ratio 3:1. The minor peaks at $m/z = 217$ and 219 are mainly the result of ^{13}C, which becomes important with this number of carbon atoms.

Fragments at $m/z = 111$ and 139 retain one ^{35}Cl atom (Figure 6.24). They differ from 216 by 105 and 77, respectively, and these fragments (if they can separately form positive ions) may well give rise to the observed peaks at

Table 6.4 *Peaks in the mass spectrum of the unknown compound Y*

m	Intensity
51	22
77	50
105	100
111	30
113	10
139	75
141	25
181	10
216	30
218	10

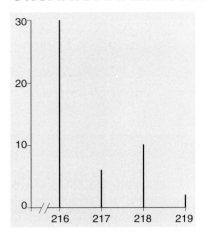

Fig. 6.23 Molecular ion region of the mass spectrum of the unknown compound Y.

these masses. $m/z = 77$ is very characteristic of a phenyl group and this is likely to be (i) the species responsible for the peak at this value and (ii) the ion fragment split from the molecular ion to give the ion of mass 139.

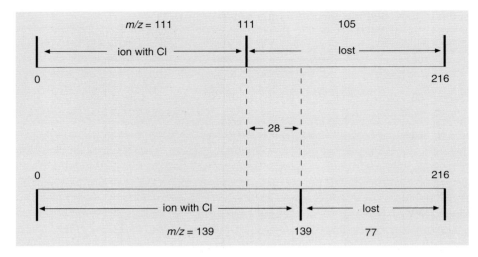

Fig. 6.24 Determining relationships between peaks.

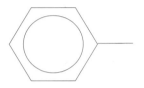

C₆H₅–

$m/z = 77$

phenyl group

The two chlorine-containing fragments differ by 28 as do the two non-chlorine-containing fragments, which indicates that the chlorine-containing group ($m/z = 111$) is joined to the phenyl group ($m/z = 77$) by a bridge of mass $m/z = 28$. Immediate possibilities are –N=N–, –CO– and –CH₂CH₂–. As the compound contains no nitrogen, –N=N– is ruled out; and as oxygen is known to be present –CO– is the only possibility. The IR spectrum would clearly show a C=O group but, taken in isolation, it may not show it to be a bridge. The smaller chlorine-containing fragment at $m/z = 111$ must contain a group of mass $111 - 35 = 76$, and if we replace the Cl by H for ease of recognition the group will have mass 77. This is likely to be a phenyl group – but not the one already identified since that could not have contained a chloro-substituent. The molecule **Y**, if known to contain one oxygen atom, must therefore be bridged by the –CO– group and have one of the structures **A–C**:

QUESTION

Assign an ion to the peak at $m/z = 181$.

A　　　　**B**　　　　**C**

The fragment at $m = 181$ is not helpful and that at 51 is presumably a ring fragment, e.g. C₄H₃. The mass spectrum does not allow distinction between these three isomers but would certainly have distinguished between, say, **D** and **E**:

QUESTION

Why must the fragment $m/z = 51$ be a ring fragment?

D　　　　**E**

Nuclear magnetic resonance spectroscopy

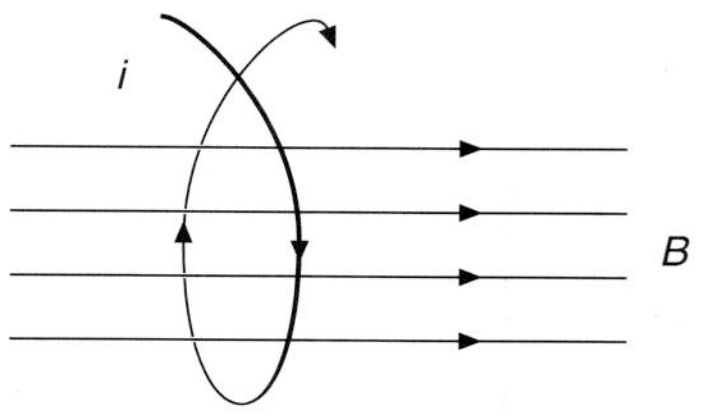

An electric current, i (one or more moving charges) in a coil has an associated magnetic field, B.

An atomic nucleus, which *must* carry an electric charge, *may* also be spinning: if it is, it will have a (very small) associated magnetic field, δB.

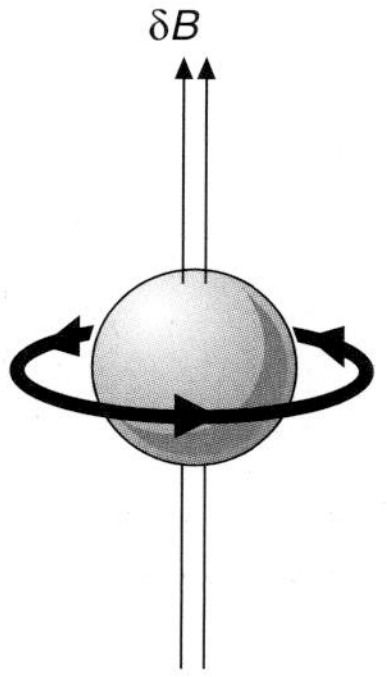

The nucleus of the common isotope of hydrogen (a proton) *does* spin and behaves as a small magnet. It is perhaps fortunate for the development of this branch of spectroscopy that the common isotopes of carbon and oxygen, ^{12}C and ^{16}O, do not.

In an external magnetic field, B, a spinning nucleus can have two or more energy levels. Again, it is fortunate that the hydrogen nucleus can have only two (Figure 6.25): a higher energy level, in which the small magnetic field of the proton is opposed to the external magnetic field, and a lower energy level, in which it is aligned with the external magnetic field. The larger the magnetic field, the larger the difference between these energy levels.

If a proton aligned with a magnetic field is supplied with electromagnetic radiation of the correct frequency to supply the energy difference between the two energy levels, it will absorb radiation and 'flip' to the higher level. This is **nuclear magnetic resonance** (Figure 6.25).

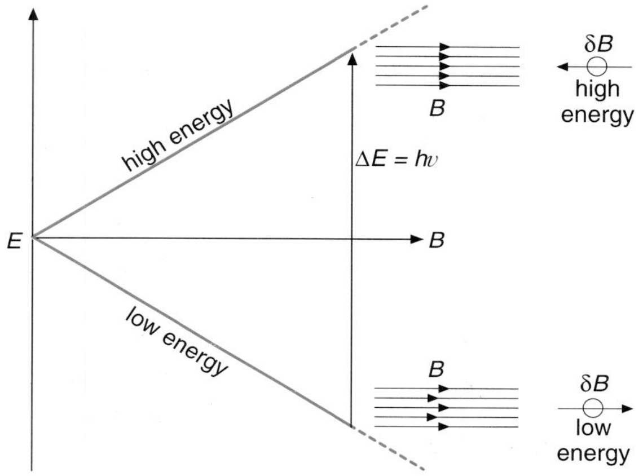

The magnetic fields have to be very large in order to make measurement of a spectrum feasible. If the field is slightly unstable or non-uniform, the spectrum will lack fine detail. Using fields of the order of 1 T (tesla), resonance of the proton occurs in the microwave region at about 40 MHz. Recent, more powerful spectrometers use superconducting coils cooled in liquid helium.

With the development of more and more sophisticated instruments, other nuclei, particularly ^{13}C, have been studied. In consequence, it has become usual to refer to nuclear magnetic resonance studies of the hydrogen nucleus as **proton magnetic resonance** (or ^1H NMR).

Fig. 6.25 The two energy levels of a hydrogen nucleus in an external magnetic field, B. As the field increases, so does the resonant frequency v.

Proton magnetic resonance

Within a molecule, local environmental effects alter the magnetic field to which a proton is subjected. The electrons in the bond holding the hydrogen atom to its neighbour interact with, and reduce, the external magnetic field. This **shielding effect** is greater the 'nearer' the bonding electrons are to the proton. They are less effective if they are shared with a strongly electronegative atom such as O in O–H or N in N–H than when shared with carbon. Hence, as the external field rises (for a given radio frequency and hence fixed ΔE), –O–H protons, which are less shielded, would resonate before –C–H protons. The standard against which the magnetic fields at resonance are measured is that for the absorption of the protons in tetramethylsilane or TMS, $(CH_3)_4Si$, which has 12 'identical' protons in the molecule.

The pure sample for examination is often dissolved in a non-proton-containing solvent such as tetrachloromethane or deuterochloroform ($CDCl_3$). The nucleus of deuterium, ^2H, does not have the same spin characteristics as ^1H and so it does not affect that part of the spectrum. A little TMS is added. Better resolution is obtained if the sample is spun in the field.

Modern spectrometers do not necessarily operate by having a fixed radio-frequency source and a varying magnetic field, or vice versa, but spectra are plotted as if this was the mode of operation.

Methanol contains protons connected to two different atoms, C and O. As the field is increased, first the –OH protons, then the methyl –CH_3 protons, and finally the TMS protons resonate in the fixed microwave radiation (Figure 6.26).

Chemical shift

The difference between the resonant field of a given proton and that of the protons in TMS is called the **chemical shift** and is given the symbol, δ. The weaker the field at which proton resonance occurs, the greater is the chemical shift and the more 'down-field' the absorption peak is said to be (see Figure 6.26). The absolute value of the chemical shift is proportional to the external field, and so is the total energy difference between the magnetic states of the proton. The shift is thus expressed as parts per million (ppm) of the field for TMS. The peak for TMS is assigned to zero on the chemical shift scale (as in Figure 6.26). The energy differences are also proportional to the microwave frequency ($\Delta E = h\nu$); hence this method of defining chemical shift also represents ppm of the oscillating electromagnetic field.

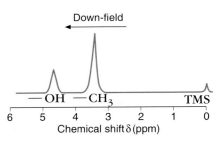

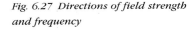

Fig. 6.26 The NMR spectrum of methanol.

On a spectrum the direction of the field strength (fixed microwave frequency) and the direction of the frequency (fixed magnetic field) run in opposite directions (Figure 6.27).

Fig. 6.27 Directions of field strength and frequency

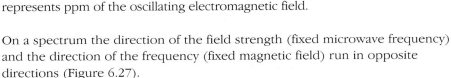

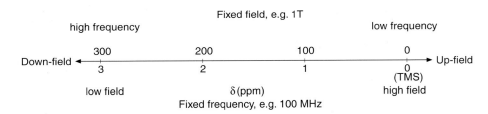

If all that NMR could do was indicate that a proportion of the hydrogen atoms was bonded to O or N rather than to C in an organic compound, it would be less useful than IR. However, the shielding effect does not require a direct connection of a proton to a different atom from carbon (often referred to as a heteroatom) in order to show itself. Protons on a carbon atom adjacent to an electron-attracting group will be **deshielded** and will be more vulnerable to the external magnetic field. They will thus resonate at a lower external field and the absorption peak will move 'down-field'. One advantage of TMS as a standard is that the silicon atom is less electronegative than carbon, a rare situation in organic compounds, and the increased shielding effect on the protons of the methyl groups pushes the absorption peak well 'up-field' and outside the range of all common absorptions.

Chemical shifts are tabulated for all groups adjacent to protons in CH, CH_2 or CH_3 and, since the *area* of the absorption peak gives a measure of the number of protons in this particular environment, the spectrum can give a far more reliable picture of a molecule than IR, but requires considerable thought in its interpretation.

Table 6.5 shows some approximate values of chemical shifts for protons in haloalkanes. Note that, as the electronegativity of the halogen increases, so does the chemical shift. The chemical shifts for protons in $-CH_2-$ and $>CH-$ are progressively greater than those for CH_3-, and the corresponding absorption is found 'down-field'.

Other electron-withdrawing groups such as C=O and O–H also deshield protons and move the resonant frequency downfield. Certain unsaturated systems, such as the benzene ring and carbonyl group when attached directly to protons, e.g. in aldehydes, exert a huge deshielding effect (as a result of interaction of their electrons with the applied magnetic field). Alkene, phenyl and aldehyde protons are likely to be found in the region $\delta = 6$, 8 and 10.

Tables of these shifts are widely published, but need treating with caution because all except terminal methyl groups have at least two groups attached, and chemical shifts cannot simply be added together.

A range of chemical shifts is shown in Figure 6.28. It may prove helpful in interpreting spectra.

Table 6.5 *Approximate values of chemical shifts for protons in haloalkanes*

Proton in	x = Cl	x = Br	x = I
CH_3X	3.0	2.7	2.2
RCH_2X	3.5	3.4	3.2

Fig. 6.28 A map of the approximate effects of single substituents (X) on the chemical shifts of protons on the same C-atom.

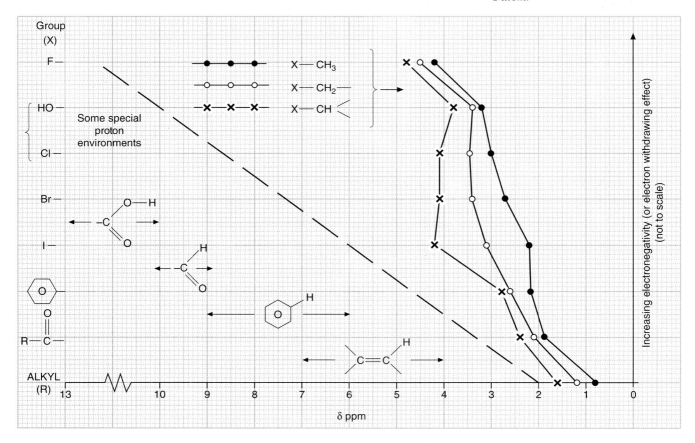

ORGANIC ANALYSIS AND STRUCTURE DETERMINATION

Spin coupling

The Edexcel syllabus requires knowledge of the application of chemical shift (and peak area) to structural determination. The importance of spin coupling is so great that it cannot be ignored. The application of low-resolution ^{1}H NMR to organic structure determination is long obsolete and you are unlikely to be able to download from the Internet, or obtain from CD ROMs or databanks, any spectra other than in high resolution. Spin coupling is therefore included here for students who wish to study chemistry or pursue the subject in a little more depth. The sample problems have been solved at both levels.

So far, we have looked at one proton in a magnetic field and considered how the external field is altered in the region of the proton by the shielding effect of the immediate bonding electrons. Nearby protons may further modify the absorption by interacting with the proton under consideration. A spinning proton influences the spin of electrons in adjacent bonds, which, in turn, influence the spin (and associated minute magnetic field) of other protons.

We shall take it as a simplifying rule that this effect is important only when protons are attached to *adjacent* (carbon) atoms.

A further simplification is that spectra are of 'first order'. This means, in effect, that coupled protons have large shift differences. We shall also assume that protons in apparently similar environments have the same chemical shift. This is not always the case but the examples have been chosen to avoid conflict.

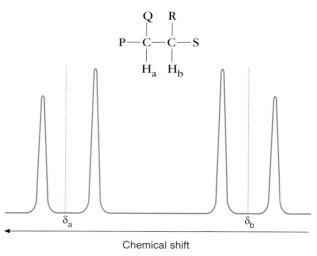

Fig. 6.29 *Each proton splits the peak of the other.*

Two protons bonded to adjacent carbon atoms in different environments might each be expected to give rise to one peak, resulting in a spectrum with two peaks. In fact, they often give rise to four such peaks (Figure 6.29), assuming that the machine can resolve them. Each proton can react to the external field in one of two ways depending on how it has been affected by the presence of the other proton.

Thus the two peaks become doublets, i.e. they are split. This effect is called **proton spin coupling**. The two peaks of the doublet are not of equal intensity unless there is a big difference in their chemical shifts. As the two proton chemical shifts become nearer, one peak of each doublet gradually dominates until, in the extreme case of two protons on the *same* carbon atom, not only does the smaller peak of each doublet disappear but also the protons both resonate at the same field and there is only one peak (of twice the area) for the pair – hence our rule about *adjacent* carbon atoms.

If a proton has n protons on *adjacent* carbon atoms, then 0, 1, ..., n of them can be aligned with it, i.e. there are $n + 1$ possible spin variations corresponding to $n + 1$ slightly different resonant fields. Thus the CH_2 peak (next to CH_3) in bromoethane is split into four tiny peaks, whereas the CH_3 peak, despite having more protons, is split into only three by the adjacent

CH$_2$ protons. The fact that there are two (CH$_2$) and three (CH$_3$) protons, respectively, is revealed by the *area* of the peaks, not by their shapes. Modern NMR machines integrate the curves and display the areas of the peaks as well as their position and shape (Figure 6.30).

It can thus be seen that many simple problems can be solved by inspection of the spectrum without a precise knowledge of δ. Provided that the chemical shift is sufficient to separate all the absorption peaks, we might expect to distinguish between the four isomers of C$_3$H$_6$Cl$_2$ by inspection of their NMR spectra (Table 6.6). Note that two terminal methyl or chloromethyl groups would give rise to only one peak, but, had we considered higher homologues than C$_3$, the identical terminal groups might have had different neighbours and would have different chemical shifts.

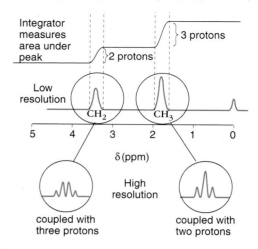

Fig. 6.30 Schematic NMR spectrum of bromoethane (CH$_3$CH$_2$Br).

Table 6.6 *Distinguishing the four isomers of C$_3$H$_6$Cl$_2$ by NMR*

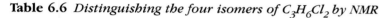

Isomer	CH$_3$CH$_2$CHCl$_2$			CH$_3$CCl$_2$CH$_3$		
Unit	CH$_3$	CH$_2$	CHCl$_2$	CH$_3$	CCl$_2$	CH$_3$
Peak type*	3	5	3	1	0	1
Area of peak	3	2	1	→†	0	→6
Total number of peaks	3			1		
Isomer	CH$_3$CHClCH$_2$Cl			ClCH$_2$CH$_2$CH$_2$Cl		
Unit	CH$_3$	CHCl	CH$_2$Cl	ClCH$_2$	CH$_2$	CH$_2$Cl
Peak type*	2	6	2	3	5	3
Area of peak	3	1	2	→	2	→4
Total number of peaks	3			2		

*Peak type: 1 = singlet, 2 = doublet, 3 = triplet, etc.

†Arrow indicates that absorption of identical groups is summed under second group. E.g. for the isomer CH$_3$CCl$_2$CH$_3$, the absorption of each methyl group is of area 3 units and both absorb at the same shift, hence the area of the single peak is 6 units.

We should be successful because three of the spectra are exactly as predicted. In the fourth, that of 1,2-dichloropropane, the two protons on –CH$_2$Cl do not behave in the same way. The difference is so slight, however, that although the expected splitting of one of the peaks is not correct, the number of peaks and their areas are as we predicted.

The coupling of protons on atoms other than carbon is complicated. Unless special precautions are taken, –O–H protons do not couple.

Hydroxy protons
In most spectra, lack of coupling gives a sharp peak. Hydrogen bonding moves the peak downfield. It is commonly found near the middle of its range of δ= 1–5 ppm.

^{13}C NMR, though technically more difficult and also more difficult to interpret, is yielding important information about the carbon skeleton of complicated molecules. NMR spectroscopy is in a state of rapid development.

ORGANIC ANALYSIS AND STRUCTURE DETERMINATION

Three problems: three ketones

On page 93 we left an uncertain situation. Although mass spectrometry had identified **C**, there was lingering doubt about **D** and **E**.

$$CH_3CH_2COCH_2CH_3 \qquad CH_3CH_2CH_2COCH_3 \qquad (CH_3)_2CHCOCH_3$$

| **C** | **D** | **E** |

The NMR spectra of **D** and **E** are given below (Figure 6.31 a and b). They are labelled NMR1 and NMR2 so that you can use them for revision. Integrator curves are not usually present on published spectra. For revision, Edexcel students should use the first two columns of numerical data in each table. Cover the rest with a sheet of paper!

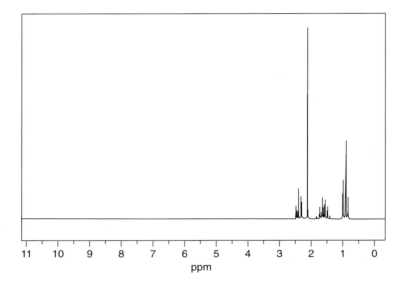

Fig. 6.31a Spectrum NMR1.

Spectrum NMR1

	δ	relative area	probable assignment	splitting type	neighbouring CH protons	comments
A	0.9	3	CH_3-	3	2	$\mathbf{CH_3}-CH_2-$
B	1.6	2	$-CH_2-$	6	5	$CH_3-\mathbf{CH_2}-CH_2-$
C	2.1	3	CH_3-	1	0	$\mathbf{CH_3}-CO *$
D	2.4	2	$-CH_2-$	3	2	$-CH_2-\mathbf{CH_2}-CO- *$

*Well down-field because of the adjacent C=O.

The structure can only be **D**. The splitting defines the neighbours completely.

Very simple structures like this can often be assigned without recourse to splitting. The spectrum shows four groups of protons. The CH_3 group protons will have different shifts because one is well downfield as a result of deshielding by the C=O group. The CH_2 group protons will similarly give rise to two peaks. This matches the number of peaks. However, since you have not defined the neighbours of each group, to complete the exercise you need either to show that at least the other two structures will not give rise to four peaks or, if you know the areas, you can show that this is further evidence. I leave the choice to you.

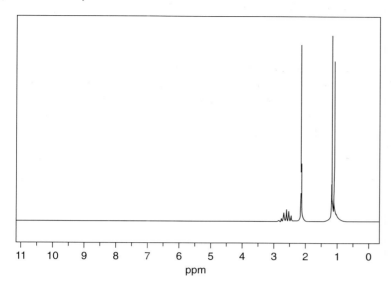

Fig. 6.31b Spectrum NMR2.

Spectrum NMR2

	δ	relative area	probable assignment	splitting type	neighbouring CH protons	comments
A	1.1	6	two CH$_3$–	2	1	**(CH$_3$)$_2$CH–**
B	2.1	3	CH$_3$–	1	0	**CH$_3$–CO** *
C	2.6	1	CH	7	6	(CH$_3$)$_2$**CH–CO** *

*Well down-field because of the adjacent C=O.

Again, the NMR spectrum defines the structure uniquely as **E**.

Using the more limited approach, we should expect **E** to have three groups of protons. The protons in the methyl group next to the carbonyl would resonate well downfield and would therefore give a different peak from those in the other methyl groups which would be expected to give one peak only. We should get two different methyl peaks and a CH peak (in the area ratio 6:3:1).

Our problem with the ketones has been solved easily with NMR.

QUESTION
Predict the NMR spectrum of pentan-2-one.

A halide problem
On page 87 we found that 1-iodopropane and 2-iodopropane had almost identical mass spectra (Figure 6.18). One of the two isomers has the NMR spectrum in Figure 6.32.

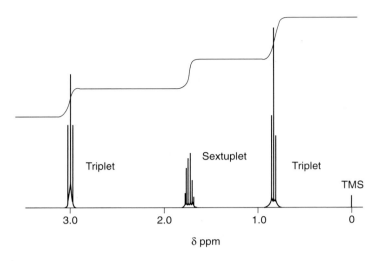

Fig. 6.32 Spectrum NMR3.

Spectrum NMR3

	δ	relative area	probable assignment	splitting type	neighbouring CH protons	comments
A	0.8	3	CH_3	3	2	**CH₃**–CH₂–
B	1.6	2	–CH_2–	6	5	CH₃–**CH₂**–CH₂–
C	3.0	2	–CH_2–	3	2	–CH₂–**CH₂**–X

The spectrum uniquely defines the structure of 1-iodopropane.

Using the simpler argument, it is not possible to define the neighbours of each CH, but for very simple structures it is not necessary. 1-Iodopropane should have three groups of peaks because there are three groups of protons with different environments. Protons in the two methylene groups will resonate at different frequencies because the one next to the iodine will be strongly deshielded and the peak will be well downfield. The areas are also consistent with one methyl and two methylene groups. It is a good idea to confirm your assignment by predicting the NMR spectrum of the other isomer to show that it would be different.

An ester problem
It can be very difficult to distinguish between isomeric esters. If the number of carbon atoms is small, this is easily done by NMR. An ester, $C_3H_6O_2$, could be ethyl methanoate or methyl ethanoate. See if you can decide which one gave rise to the spectrum shown in Figure 6.33.

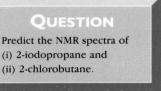

QUESTION

Predict the NMR spectra of
(i) 2-iodopropane and
(ii) 2-chlorobutane.

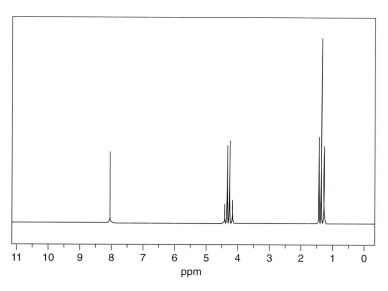

Fig. 6.33 Spectrum NMR4.

Spectrum NMR4

	δ	relative area	probable assignment	splitting type	neighbouring CH protons	comments
A	8.0	1	CH	1	0	**H–CO**–O *
B	4.2	2	–CH$_2$–	4	3	CH$_3$–**CH$_2$**–O **
C	1.3	3	CH$_3$–	3	2	**CH$_3$**–CH$_2$–

* This isolated (no splitting) proton must be in a –CHO group with such a shift.

** So far down-field it must be next to O or Cl; O is the only atom available.

The relative areas might not be available, but the splitting pattern defines the structure uniquely.

The simpler approach looks at the number of peaks — in the case of ethyl methanoate this is clearly three. If you can assign them by area then there is no need to use coupling. Alternatively, you can show that the number of peaks in the spectrum of methyl ethanoate would be different.

Ultraviolet and visible spectra

The easily usable UV region stretches from about 200 to 400 nm wavelength. The visible region is from about 400 to 750 nm.

UV and visible spectra are the result of electronic transitions involving much greater energy changes than those met in IR and NMR. The exact nature of the transitions is not significant at this level but you should realise that only the absorption associated with π-systems in C=C, C=O or benzene is accessible with normal laboratory equipment. These absorbing groups are called **chromophores**.

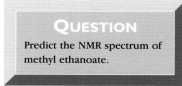

QUESTION

Predict the NMR spectrum of methyl ethanoate.

The spectrum is plotted as the strength of the absorption, expressed as an extinction coefficient, ε, against the wavelength, λ, in nanometres, nm (Figure 6.34a and 6.34b. ε is calculated from the ratio of incident to transmitted light, the size of the cell used to hold the solution and the concentration of the solution. It is often a large number and may be plotted as log ε. By using the logarithm small peaks are not lost. If two peaks had ε values of 10 000 and 10 respectively, then plotting them directly would give a ratio of 1000:1 and the small peak would be invisible. Using $\log_{10}$ values, the ratio would be 4:1 and both peaks would easily be seen. The spectra are pretty featureless and printing them out takes up space to little advantage. It is thus the practice to publish only the values of the wavelength of the maximum absorption (there may be only one peak), λ_{max}, and the corresponding value of ε.

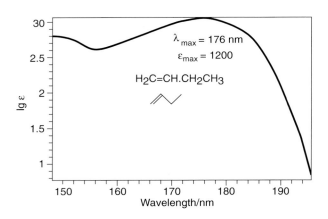

Fig. 6.34a UV spectrum of 1-butene.

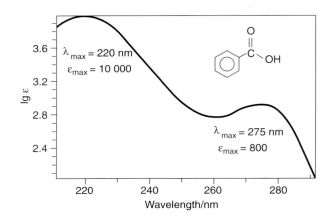

Fig. 6.34b UV spectrum of benzoic acid.

The broad peaks (perhaps 'humps' would be more appropriate) are by no means as diagnostic of structure as they are in IR and NMR spectra. Often, when more than one chromophore is present, one peak tends to bury another. The main use of UV is in the detection of conjugated systems of double bonds. These are at least two double bonds alternating with single bonds.

Conjugated	Isolated
$-CH=CH-CH=CH-CH=CH-$	$-CH=CH-CH_2-CH_2-CH=CH-$
$-CH=CH-C=O$	$-CH=CH-CH_2-C=O$
$\quad\quad\quad\quad\mid$	$\quad\quad\quad\quad\mid$

Conjugation has the effect of moving the absorption to longer wavelengths and increasing the strength of the absorption – 'longer and stronger'. Compare (Table 6.7) the absorption of the isolated chromophore in butene with that of the conjugated system in butadiene, $H_2C=CH-CH=CH_2$, and the extended system in benzoic acid, C_6H_5-COOH with the even more extended one in 3-phenylpropenoic acid, $C_6H_5-CH=CH-COOH$. Incidentally, when the main band of 3-phenylpropenoic acid 'moved' from $\lambda_{max} = 220$ nm in benzoic acid to 275 nm, with a large increase in strength, it completely swallowed up the smaller peak.

Table 6.7 *Position of main absorptions of butadiene and 3-phenylpropenoic acid*

Compound	λ_{max} / nm	ε_{max}
butadiene	217	21 000
3-phenylpropenoic acid	275	21 000

Suppose an unsaturated compound, **A**, C_5H_9Br, were treated with hot, ethanolic potassium hydroxide to give a pentadiene, **B**, C_5H_8, with the UV spectrum below.

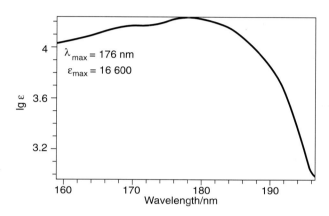

λ max = 176 nm
ε max = 16 600

Fig. 6.35 UV spectrum of pentadiene B.

Comparing λ_{max}, and ε. with butene and butadiene shows that the double bonds of **B** must be isolated. There is only one possibility for **B**, $H_2C=CHCH_2CH=CH_2$ and of the two possibilities for **A**, $H_2C=CHCH_2CH_2CH_2Br$ (**A1**) and $H_2C=CHCH_2CHBrCH_3$, (**A2**), only **A1** is likely, since **A2** would have given some, if not all, of the conjugated species.

Only when extended conjugation allows considerable delocalisation (e.g., methyl orange), does the absorption move into the visible spectrum.

Yellow form of methyl orange

The use of UV and visible spectra in structure determination is extremely limited. The main use today is in the quantitative estimation of amounts of materials in samples by dissolving them and comparing their light absorption with known samples at the same wavelength. This is often extended into inorganic chemistry, e.g. the manganese content of a solution could be found

QUESTION

How would you intensify the colour of a copper(II) salt solution in order to make it suitable for colorimetry?

by oxidation to manganate(VII) (purple) and comparing the colour intensity (light absorption) with that of known solutions. If this were a routine operation, a spectrophotometer or colorimeter might well be calibrated for that purpose. A wide range of reagents form highly coloured complexes with metals and are used for this purpose.

Questions

Nuclear magnetic resonance

You may find Figure 6.28 helpful.

1 (*a*)

cyclooctatetraene   phenylethene

The low resolution NMR spectra of these two compounds are sketched below. Which is which?

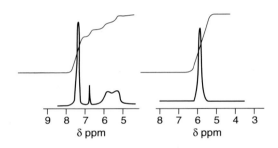

(*b*) Predict the low resolution NMR spectrum of the anaesthetic cyclopropane C_3H_6.

2 Predict low resolution NMR spectra of the four isomers of C_4H_9Cl. Use them to identify the isomers with the spectra below.

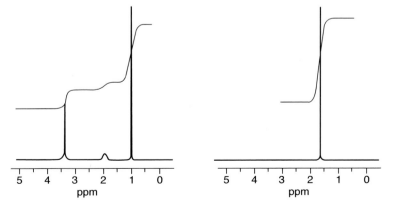

3 The NMR spectrum below is that of a simple monohydric alcohol. Identify it and explain as much of the spectrum as you can.

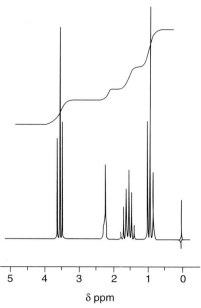

δ ppm

Infra-red/ultraviolet spectra

You will need to refer to a table of IR absorption frequencies (wavenumbers), such as that on p. 83.

4 An unbranched compound, A, gives a positive iodoform test and gives a molecular ion peak at $m/z = 86$ in the mass spectrum. The infrared spectrum of A is shown below.

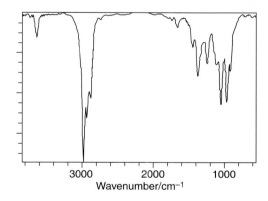

Wavenumber/cm^{-1}

Treatment of A with ethanoyl choride gave B. Predict two differences you would see in the infrared spectra of A and B.

5 A neutral compound, N, has the infra-red and mass spectra given below.

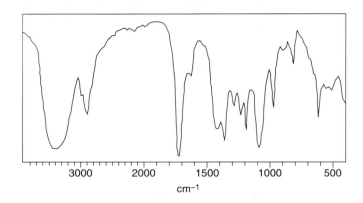

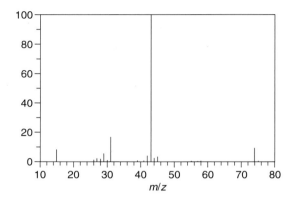

Deduce the structure of N.

Mass spectra

6 A pleasant smelling liquid, L, gives a red precipitate with 2,4-dinitro-phenylhydrazine reagent. The mass spectrum of L is given below.

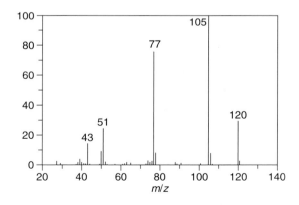

Deduce the structure of L.

7 A pleasant smelling compound A, on boiling with alkaline potassium manganate(VII) gives (after acidification) a crystalline solid B, $C_7H_6O_2$. The mass spectrum of A is given below.

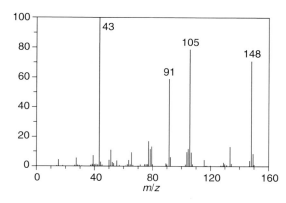

Deduce the structure of A, assign ions to the four main peaks and comment on the general appearance of the spectrum.

8 One of a pair of geometrical isomers, C, had the proton NMR spectrum below. On treatment with hydriodic acid it gave only one product, D, the mass spectrum of which is given.

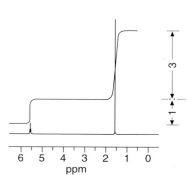 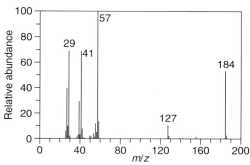

Give the structure of D, and of the two geometrical isomers, one of which is C. Support your choice of structures with chemical and spectral evidence.

9 An organic compound, E, $C_pH_qX_r$, readily gives a pale yellow precipitate with ethanolic silver nitrate. E has the mass spectrum below.

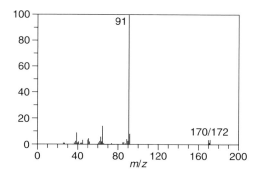

(i) Identify X; (ii) give the values of p, q and r; (iii) give the structure of E. Explain your reasoning.

10 A compound, F, of C, H and Br, did not give any reaction with ethanolic silver nitrate. Deduce what you can about its structure from the mass spectrum below.

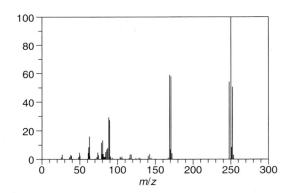

Explain the significance of the groups of ions centred on $m/z = 170$ and 250.

Organic synthesis

You can make anything from anything

From the point of view of the organic chemist, this is probably true, but it would be a foolhardy chemist who tried to make gold from hydrogen. The latter change might be feasible in a supernova, but it certainly is *not* a sound proposition in a chemistry laboratory. Even for the organic chemist, the starting point is never just 'any old thing': it is carefully chosen.

Choice of starting material and route

In reality, the organic chemist tries to make the target compound from the nearest structure that is commercially available. This is often a natural product, e.g. the manufacture of codeine from morphine.

QUESTION

An expensive material, A, is to be converted into B. Route 1 is one step with 45% yield. Route 2 is two steps, each with 60% yield. Which is better?

The route chosen is normally the shortest possible. **Organic reactions never give 100% yields**. If, on average, each stage of a reaction sequence gave a 50% yield (which would by no means be untypical) and ten stages were required, the final yield would be $(\frac{1}{2})^{10}$, i.e. 1/1024 or 0.1%, of the theoretical: grams from kilograms! A five-stage process, perhaps using more expensive reagents (e.g. lithium tetrahydridoaluminate(III)), on the same 50% yield basis, would give $(\frac{1}{2})^5$, or about 30 times as much as the ten-stage alternative.

Compare two possible routes from propanol to butanol. It is very likely that any pathway would begin by preparing bromopropane:

$$C_3H_7OH + KBr + H_2SO_4 \rightarrow C_3H_7Br + KHSO_4 + H_2O$$

Two alternatives (at least) are then possible:

- Route A

$$C_3H_7Br \xrightarrow{\text{step 1}} C_3H_7CN \xrightarrow{\text{step 2}} C_3H_7CH_2NH_2 \xrightarrow{\text{step 3}} C_4H_9OH$$

  Step 1: ethanolic potassium or sodium cyanide gives a good yield, but the highly toxic reagent is an obvious hazard.

  Step 2: lithium tetrahydridoaluminate(III) again gives a good yield but is a very expensive reagent.

  Step 3: sodium nitrite and hydrochloric acid gives a very poor yield of butanol.

- Route B

$$C_3H_7Br \xrightarrow{\text{step 1}} C_3H_7MgBr \xrightarrow{\text{step 2}} C_4H_9OH$$

Step 1: magnesium and ethoxyethane ('ether' solvent) gives an excellent yield of the Grignard reagent, which is used *in situ*.

Step 2: treatment with methanal followed, *in situ*, by decomposition with aqueous acid gives a good yield of the product.

Route A uses an unpleasant reagent and an expensive reagent; it involves three steps and gives a poor yield. Route B is, effectively, one step because the whole sequence is carried out without isolating an intermediate, and it gives a good yield. Route B would be the obvious choice.

Old A-Level papers are full of questions that ask the candidate to make **B** from **A** *without the use of any other organic compounds* (except perhaps solvents), e.g. ethyl ethanoate from ethanol:

$$CH_3CH_2OH \overset{K_2Cr_2O_7/H^+}{\underset{conc\ H_2SO_4}{\rightleftharpoons}} CH_3CO_2H \overset{ethanol}{\rightleftharpoons} CH_3CO_2CH_2CH_3$$

The restriction, here, is to make sure that the candidate does not start by using ethanoic acid but must make it first.

The organic chemist is not fettered by the pedantic demands of the examiner. The best way is the fastest, or the most high yielding, or the cheapest.

The inclusion of Grignard reagents in the A-Level specification means that there is more opportunity to ask questions in which more than one organic material may be required at some stage. For example, how would you convert propan-2-ol into 2-methylpropan-2-ol? A possible answer is shown in Figure 7.1. Here a Grignard reagent is involved, and it would be rather silly to expect anyone to make it from the propan-2-ol.

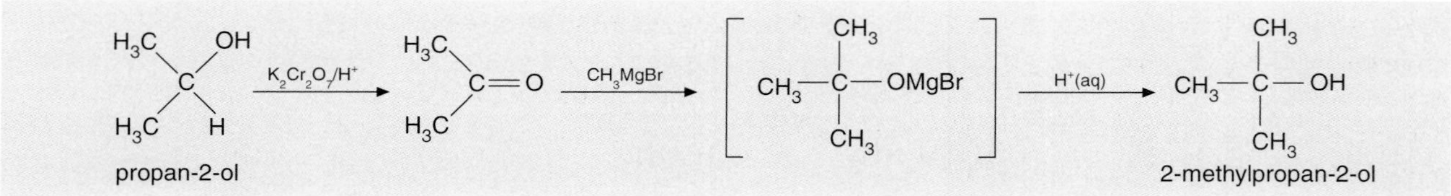

Fig. 7.1 Reaction scheme to convert propan-2-ol into 2-methylpropan-2-ol.

QUESTION

What products would be formed if you used:
(i) butan-2-ol at the start instead of propan-2-ol;
(ii) ethylmagnesium bromide as the Grignard reagent;
(iii) both of these?

But, even in questions involving Grignard reagents, the old restriction could still be applied. For example, how would you make hexan-3-ol from propan-1-ol using no other organic reagents except solvents? A possible answer is shown in Figure 7.2.

It is important to realise that the organic reactions in the A-Level syllabus are, with the exception of the use of lithium tetrahydridoaluminate(III), all (fairly unsophisticated) nineteenth-century methods. They represent an infinitesimally small fraction of all reactions in use today. Nevertheless, their use illustrates many general principles, e.g. the idea that a group that appears to need retaining may first have to be destroyed and then later replaced. This is the case in the two

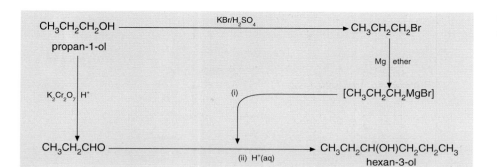

Fig. 7.2 Reaction scheme to convert propan-1-ol into hexan-3-ol.

examples involving Grignard reagents in Figures 7.1 and 7.2, and for the –OH group in the preparation of 2-hydroxy-2-methylpropanenitrile (**B**) from propan-2-ol (**A**) in Figure 7.3. The inexperienced student is likely to look for a way of directly replacing the 2° hydrogen by a cyanide when faced with this problem.

*Fig. 7.3 Reaction scheme to convert propan-2-ol (**A**) into 2-hydroxy-2-methylpropanenitrile (**B**).*

Main groups of synthetic reactions

Synthetic procedures fall roughly into four groups, depending on whether the number of carbon atoms in the carbon skeleton of the molecule is changed:

(i)   no change to carbon skeleton;
(ii)   number of carbon atoms in skeleton increases;
(iii)   number of carbon atoms in skeleton decreases by one;
(iv)   polymerisation occurs

One of the first decisions that must be taken when planning a route is whether such a change is required. The reference to the carbon *skeleton* is necessary because reactions like esterification or ester hydrolysis, while changing the number of carbon atoms in the molecule, do not change the size of the carbon skeleton.

No change in the carbon skeleton
Reactions of this type include:

• simple substitution reactions involving non-carbon-containing groups, e.g. –OH, –Br, –I, –NH$_2$, etc.

• limited oxidation of alcohols and aldehydes, or reduction of aldehydes, ketones, nitriles and carboxylic acids

• eliminations yielding C=C double bonds, amides and nitriles, ester hydrolysis and esterification

These reactions are usually a necessary preliminary in order to reach a point where the number of carbon atoms can be changed. Some of the more common situations are illustrated in Figure 7.4.

ORGANIC SYNTHESIS

Fig. 7.4 Some reactions where there is no change in the carbon skeleton.

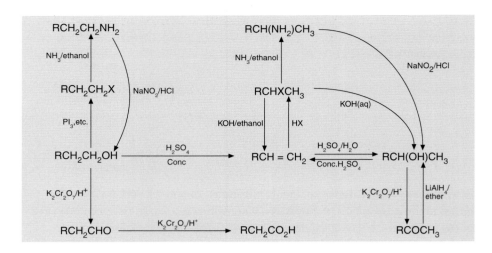

QUESTION

How would you convert 1-bromopropane into 2-bromopropane?

QUESTION

Give as many synthetic reactions as you can where there is no change in the carbon skeleton. Then look at Figure 7.4.

QUESTION

How would you convert propan-2-ol into
(i) 2,3-dimethylbutan-2-ol;
(ii) propene?

QUESTION

Which important method of extending the carbon skeleton is missing from Figure 7.5?

Number of carbon atoms in the skeleton is increased

... by one

Such reactions include:

- introduction of the cyanide group by nucleophilic substitution of a haloalkane, or addition of HCN to a carbonyl compound

- introduction of the primary alcohol –CH₂OH or carboxylic acid groups by addition of methanal, HCHO, or carbon dioxide to a Grignard reagent

... by more than one

Such reactions include:

- addition of Grignard reagents to aldehydes and ketones

- Friedel–Crafts additions to aromatic systems

Some of these reactions are shown in Figure 7.5.

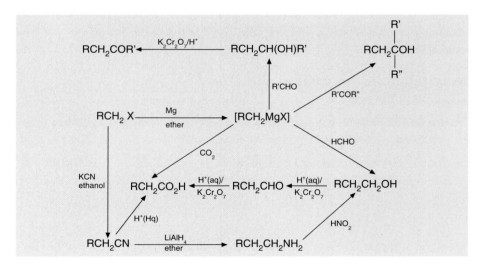

Fig. 7.5 Some reactions where there is an increase in the number of carbon atoms in the skeleton.

Number of carbon atoms in the skeleton is decreased by one (degradations)

These methods include:

- the Hofmann degradation of amides with sodium hydroxide and bromine

$$RCONH_2 \; + \; NaOBr \; \rightarrow \; RNH_2 \; + \; NaBr \; + \; CO_2$$

- the oxidation of aromatic side-chains with alkaline potassium manganate(VII)

$$C_6H_5CH_3 \; \overset{KMnO_4/OH^-}{\rightarrow} \; C_6H_5CO_2H$$

- the haloform reaction, applied to methyl ketones and methyl secondary alcohols

$$RCH(OH)CH_3 \; \overset{X_2/NaOH}{\rightarrow} \; RCOCH_3 \; \overset{X_2/NaOH}{\rightarrow} \; RCO_2H$$

where X is a halogen; this is iodine if the reaction is to be used as a test for the presence of the CH_3CO- or $CH_3CH(OH)-$ groups.

Polymerisation reactions

These reactions have been dealt with fully in Chapter 3. The only point to add here is that examination of the repeat unit will indicate whether polyaddition or polycondensation is required. If the chain is an unbroken sequence of carbon atoms, then an alkene must be made to undergo polyaddition.

Most of the condensation polymers are either polyesters or polyamides, which involve the joining of molecules with two functional groups. Some care is needed in examining a polymer if the repeat unit is not given. Thus

$$—CH_2CH_2CH_2CH_2CONHCH_2CH_2CH_2CH_2CONHCH_2CH_2CH_2CH_2CONH—$$

has the repeat unit

$$—NHCH_2CH_2CH_2CH_2CO—$$

and would be made by the polycondensation (probably by heat) of the single amino acid

$$H_2NCH_2CH_2CH_2CH_2CO_2H$$

whereas the similar, but not identical, polyamide

$$—CH_2CH_2CH_2CH_2CONHCH_2CH_2CH_2CH_2NHCOCH_2CH_2CH_2CH_2CONH—$$

has the repeat unit

$$—NHCH_2CH_2CH_2CH_2NHCOCH_2CH_2CH_2CH_2CO—$$

QUESTION

How would you convert phenylethanone into
(i) benzoic acid;
(ii) phenylamine;
(iii) 2,4,6-tribromophenol
(in that order)?

QUESTION

What does the carbon atom of the methyl group become in the haloform reaction?

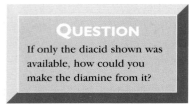

QUESTION

If only the diacid shown was available, how could you make the diamine from it?

and would need to be made by the co-polymerisation of the following diamine and dicarboxylic acid:

$$H_2NCH_2CH_2CH_2CH_2NH_2 \qquad HO_2CCH_2CH_2CH_2CH_2CO_2H$$

Similar care must be used when examining the structure of polyesters. They, too, can be made by the polycondensation of a hydroxy acid, or, like terylene, by the co-polymerisation of a diol and a dicarboxylic acid.

Practical synthetic techniques

In *elementary* synthesis, the variety of methods is small. Most reaction techniques are concerned with mixing and temperature control.

Mixing

Unlike the reactions of inorganic chemistry, those of organic chemistry often use immiscible liquid reactants, and it is thus necessary to shake or stir the mixture.

Shaking is normally only practicable in stoppered flasks for reactions that do not build up pressure or in flasks fitted with condensers for reflux (see below). Sometimes, when there is an appreciable difference in the densities of immiscible liquids, e.g. in the nitration of benzene, or when, as in the reduction of nitrobenzene to phenylamine, a dense solid is involved as well, vigorous shaking is the only feasible method of bringing the reactants together adequately.

A mechanical stirrer is normally used if the reaction takes an appreciable time. This may take the form of a motor-driven 'paddle', which dips into the reaction mixture, or, for reactions that do not need heating, a magnetic stirrer may be employed (Figure 7.6). The latter is simply a small iron bar coated with an unreactive plastic material (to prevent reaction with the iron or damage to the glass), which is spun round under the influence of a magnet that is rotated below the flask.

Temperature control

Compared with inorganic reactions, which are often ionic and very fast, most organic reactions are relatively slow. The formation of the transition states usually involves breaking of covalent bonds (homolytically or heterolytically), for which high activation energies are needed. Hence the reactions are often heated. Reactions that begin vigorously, e.g. the nitration of benzene or the reduction of nitrobenzene, often require heating to bring about completion. For safety, this is usually done on a steam bath or by using an electric heating mantle (see Figure 7.7), but the heating brings another problem. Organic materials are usually volatile, to some extent, and heating the mixture would cause them to escape. Apart from the consequent reduction in yield, this has associated fire and health hazards. In order to prevent loss of materials, the heating is usually done under **reflux**. Here a condenser is fitted, often vertically, to the reaction flask so that condensed vapour is returned directly to the flask.

Fig. 7.6 A magnetic stirrer in use.

Cooling is often required, usually to prevent reactions becoming uncontrollable, e.g. the nitration of a compound such as cellulose can suddenly become an explosive oxidation if the heat of reaction is allowed to raise the temperature appreciably. In some reactions, temperature control is needed to prevent further reaction, e.g. the production of excessive amounts of dinitrobenzene when nitrating benzene. Such cooling is usually achieved by placing the reaction flask under a stream of cold water over the sink.

Filtration

In preparative inorganic chemistry, filtration is usually employed to remove small amounts of *unwanted* material from a liquid, e.g. undissolved solid from a solution or excess of a solid reactant from a liquid mixture. Recovery of the solid is not normally important and it is spread thinly over the surface of a large filter paper so as to give the fastest filtration. In organic chemistry, the solid is often the required product, and spreading it thinly would give large **mechanical losses**, a term used to describe material lost during physical operations. Here much material would be irrecoverable from the surface of the paper. However, a thick deposit of solid in a filter slows filtration down to an unacceptable rate.

The problem is overcome by using a **Büchner funnel** (Figure 7.8), or similar device, in which reduction of pressure below a perforated plate supporting the filter paper causes atmospheric pressure to drive the liquid through. Not only does this speed up the process, but also the pad of solid on the filter has much of the included liquid forced down into the receiving flask at the end. Modern funnels of this type are often made in three pieces: this assists in the removal of the cake of solid and makes cleaning of the funnel easier.

It is important, during the formation of the solid, to allow it to form slowly; e.g. if the solid crystallises, the solution should be allowed to cool slowly. The larger particles so produced do not clog the filter. It is also important not to evacuate the receiving flask too strongly, especially at the start of filtration, because the unsupported areas of wet filter paper, over the perforations, can collapse. Trapped solvent or reaction mixture in the solid will be removed by recrystallisation.

Fig. 7.7 Boiling under reflux can be carried out safely using a heating mantle.

Fig. 7.8 A Büchner funnel in use and 'in pieces'.

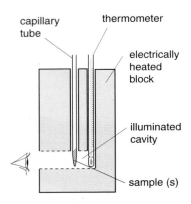

Fig. 7.9 Melting point apparatus.

> ## QUESTION
>
> Why is it helpful, when purifying by recrystallisation, if an impurity is much less or much more soluble than the desired compound in the chosen solvent?

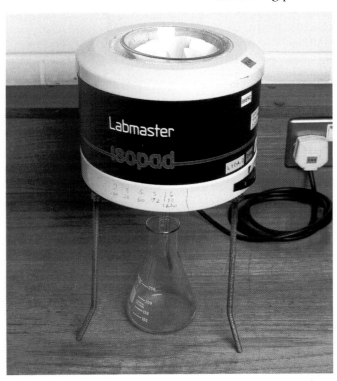

Fig. 7.10 A heated funnel and a fluted filter paper.

Melting point

Purity of a solid is often judged by its melting point. Each time a solid product is recrystallised, a melting point is measured. The most common melting point apparatus in schools consists of an electrically heated block drilled to take a thermometer and up to three capillary tubes for specimens (Figure 7.9). The temperature is slowly raised until the content of the capillary tube, viewed through a lens and illuminated is seen to melt.

When purified by recrystallisation, a small sample is crushed on a clean porous tile in order to put it in the capillary tube. The tile absorbs any traces of solvent. The melting point should be sharp or the sample is not pure. The temperature of the heating block must be raised very slowly – about 3°C per minute near the melting point. If the temperature rises while the sample is melting, not only will a 'high' result be obtained but also the sample will appear to melt over a temperature range and seem to be impure.

In the days when this technique was used to show that an unidentified solid, X, was identical with a compound Y, the melting points of samples of X, the known Y, and a mixture of X and Y, were all determined simultaneously (hence three tubes). Ideally, all should melt together.

Recrystallisation of solids

This is extensively used for the purification of solids and is loosely used whether the solid has been previously crystallised or not. A suitable solvent must not react with the solid. The solubility of the solid should be high near the boiling point of the solvent and as low as possible at room temperature. If possible, the impurities should either be insoluble in it (unlikely) or very soluble in it. The solvent should not normally boil at a temperature above the melting point of the solid, since the solid may be initially precipitated as a liquid. This will then precipitate in lumps, trapping solvent and impurities (dissolved in the solvent).

The compound for purification is dissolved in the minimum of hot solvent. The hot solution is filtered to remove insoluble matter. The solution may cool and cause premature precipitation, but the problem can usually be overcome by using a slight excess of solvent, a pre-heated funnel and a fluted filter paper (folded for fast filtering). Funnels with heating jackets can also be used (Figure 7.10). The hot solution is then allowed to crystallise in a conical flask with a loose stopper to keep out dust and dirt.

The crystalline solid is usually separated from the cold solvent by filtration using a Büchner funnel. There is always appreciable loss of the solid because of its solubility. This can be minimised by finally cooling the solution in a refrigerator if the impurities are likely to be in insufficient amount to saturate the solution at this temperature. When filtering, the remaining solution must be removed from the surface of the crystals or it will evaporate, leaving dissolved impurities. For this reason, it is usual to wash the solid on the filter with a *little*, sometimes ice-cold, solvent.

Recrystallisation is usually repeated until the material has a constant, sharp melting point.

If a solid is discoloured, recrystallisation often fails to remove all of the impurity responsible. Treating the hot solution to be crystallised with a little activated charcoal before hot filtration often removes the coloured impurity by adsorption. It will doubtless remove other impurities too, but it will also cause some loss of the desired product.

Simple distillation

If the product is sufficiently different in boiling point from the rest of the reaction mixture, and does not form an **azeotrope**, simple distillation may be used. Thus, bromoethane may be prepared by the reaction of potassium bromide, ethanol, concentrated sulphuric acid and a little water. The boiling point of bromoethane (38°C) is very different from that of ethanol (78°C), water (100°C) and concentrated sulphuric acid (>300°C). Simple distillation removes the crude compound from the mixture, but it will be contaminated by small amounts of ethoxyethane (bp 35°C), which is always formed in this reaction, and by dissolved hydrogen bromide and perhaps a trace of volatile bromine.

Fractional distillation

Organic chemists usually separate mixtures of liquids by the process of fractional distillation. This involves the use of a fractionating column. A typical laboratory apparatus is shown in Figure 7.12, but the design and complexity of distillation (fractionating) columns varies greatly. Those for the fractionation of liquid air separate the components, but this low-temperature distillation is usually carried out in two stages under different pressures. Petroleum has many components: the commercial distillation simply vaporizes all that will boil and continuously removes *mixtures*, which are 'fractions' with a range of boiling points from various condensation trays at different heights of the column.

It is tempting to suppose that, if we boil a mixture of liquids, they will leave the mixture as the temperature rises, one at a time, in order of increasing boiling point. Such is not the case. It may be possible to separate all of them by **fractional distillation**, but not because they leave the liquid in this simple order.

Suppose that we tried to separate an ideal binary liquid mixture in a sequence of simple distillations. Suppose that we boil a great deal of the mixture of composition c_1 (Figure 7.13). It will boil at temperature t_1 and the vapour in equilibrium has the composition c_2 (richer in the more volatile **B**). Of course, we cannot boil off much vapour before affecting the composition of the liquid, which must be getting richer in the less volatile **A**; and if we do affect the liquid composition, that in turn will affect the vapour. We condense *all* of the small amount of vapour – that way its composition cannot change because it is not in equilibrium – and we have a very small amount of distillate of unchanged composition c_2 (Figure 7.13).

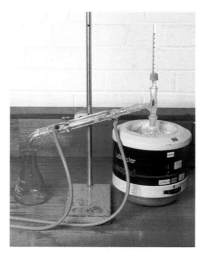

Fig. 7.11 Simple distillation in the laboratory.

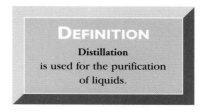

DEFINITION
Distillation
is used for the purification
of liquids.

Fig. 7.12 A laboratory fractional distillation apparatus.

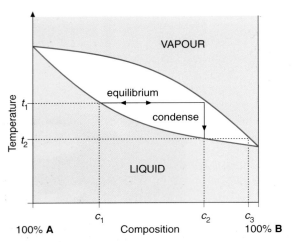

Fig. 7.13 Phase diagram for the separation of an ideal binary mixture.

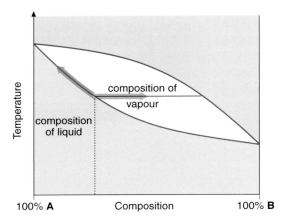

Fig. 7.14 Phase diagram for an ideal binary mixture showing the direction of continuous change in composition of liquid and vapour.

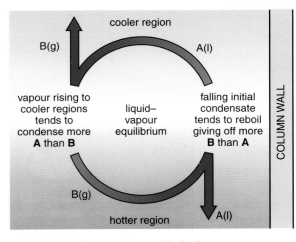

Fig. 7.15 Equilibrium at the wall of a fractionating column.

We then boil the small amount of liquid of composition c_2 (in fresh apparatus). It boils at temperature t_2 and the vapour in equilibrium will have composition c_3. But, of course, we can only take a minute amount of the vapour this time without affecting the composition of the liquid, and if we distilled all of the liquid we would achieve no purification whatsoever! Finally, by repetition of the process, as can be seen from Figure 7.13, we should be able to get a pathetically small amount of (essentially) pure **B**.

Such a stepped diagram shows that the separation of **B** from **A** is theoretically possible, and it can be used by chemists to estimate the difficulty of separation of a particular mixture (the more 'steps' the harder it will be). It is also used to estimate the efficiency of distillation columns based on the number of 'steps' achieved in one distillation using the equipment under test.

It is not very satisfactory to explain fractional distillation in this simple and pedestrian way, but it does illustrate that the process is possible.

The operation of a fractionating column involves a temperature gradient up the column and a corresponding concentration or composition gradient (Figure 7.14). At any moment of time, an infinite number of equilibria between vapour and liquid exist in the column.

As soon as the liquid in the flask boils and gives off vapour richer in **B**, the liquid becomes richer in **A** and its boiling point rises. Further vapour that leaves the liquid does so at a progressively higher temperature and, although the vapour will always be richer in **B** than the liquid in equilibrium, the percentage of **B** in both is continually falling. The first vapour condenses, thereby heating the column, and runs back down the column to be reboiled. Gradually the column develops a temperature gradient and continuously brings descending condensate into equilibrium with rising vapour (Figure 7.15).

An efficient column is either long enough or sufficiently sophisticated in its design to bring about the necessary equilibration between evaporate and condensate, so that, in operation, the top of the column is approximately at the boiling point of the more volatile component (**B**) and this distils.

Some liquid mixtures cannot be separated entirely by distillation because they form azeotropes. Students who wish to know more about these should look in a book of physical chemistry or the first edition of this series.

Boiling point determination

The best way to determine a boiling point is to carry out a distillation and observe the steady temperature when this occurs. If, however, the sample is small and the boiling point is required it may be determined by Siwolobov's method. This, by modern standards, requires a lot of material (up to 0.5 cm^3) and compound identification by boiling point has long been superseded by the use of spectra. The sample is placed in a wide, thin-walled capillary tube, into which is placed a narrow capillary tube sealed at the upper end with its mouth below the surface. The whole is fastened to a thermometer with the liquid next to the bulb. This is placed in an oil bath and the temperature is raised slowly (see melting point on p. 118). Air slowly escapes from the smaller capillary until, at the boiling point of the sample, a rapid stream of bubbles occurs. Many consider that a better estimate of the boiling point is obtained by allowing the oil bath to cool and recording the temperature when the liquid first rises in the inner capillary, but a larger sample is required because of evaporation.

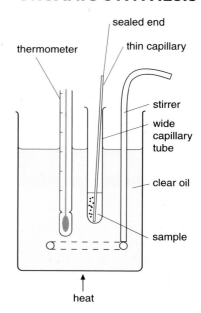

Fig. 7.16 *Boiling point determination by Siwolobov's method.*

Questions

1 How would you prepare from propene:

   (i) 2-bromopropane;

   (ii) propanone;

   (iii) 2,3-dimethylbutan-2-ol?

2 How would you make

   (i) phenol,

   (ii) benzoic acid,

   (iii) phenylbenzoate,

using benzene and methylbenzene as your only organic materials?

3 How would you convert oleic acid $CH_3(CH_2)_7CH=CH(CH_2)_7CO_2H$ into

   (i) stearic acid, $CH_3(CH_2)_{16}CO_2H$;

   (ii) palmitic acid, $CH_3(CH_2)_{14}CO_2H$;

   (iii) octadeca-8,10-dienoic acid,
       $CH_3(CH_2)_6CH=CHCH=CH(CH_2)_6CO_2H$?

4 Starting from butan-1-ol and methanol as your only organic materials, how would you make

   (i) butanone;

   (ii) propanoic acid;

   (iii) methyl butanoate;

   (iv) propylamine;

   (v) butene;

   (vi) 2-methylbutene?

8 The commercial importance of organic compounds

Organic chemistry influences every walk of life. The clothes we wear are often blends of wool or cotton with synthetic fibres such as nylon or terylene. The shoe-repairer, once a feature of every town and village, has all but disappeared with the marketing of 'throw-away' footwear, shoes with hard-wearing synthetic soles and often synthetic uppers as well, probably glued together with a synthetic adhesive.

Polymers have had consequences never dreamt of by Leo Baekeland, inventor of the first modern 'plastic', Bakelite, early in the twentieth century. Their uses range from 'Lego' sets to the interiors of jet air-liners. Some are shown in Figure 8.1.

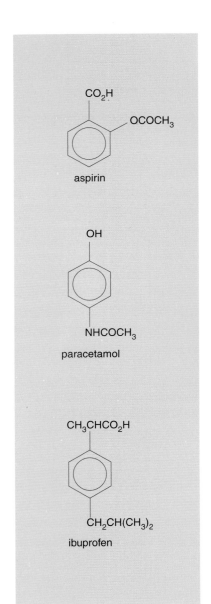

aspirin

paracetamol

ibuprofen

Fig. 8.2 The structures of aspirin, paracetamol and ibuprofen.

Fig. 8.1 Some uses of modern plastics.

Pharmaceuticals

Two distinct approaches are applied to the synthesis of drugs. Simple drugs like the analgesics aspirin, paracetamol and ibuprofen (Figure 8.2) are purely synthetic. More elaborate pain-killers like morphine and codeine (Figure 8.3), antibiotics like penicillin, or physiologically active hormones are made by biosynthesis and subsequent modification. However, the great benefits of the vast range of modern antibiotics and drugs have been accompanied by the social problems of drug abuse.

THE COMMERCIAL IMPORTANCE OF ORGANIC COMPOUNDS

Morphine is extracted from poppy seeds, codeine is made by the chemical modification of morphine, and penicillins are made (initially) by mould culture. Great progress is being made in the manufacture of materials like the hormone insulin. Traditionally extracted from an animal pancreas, it can now be made by genetically engineered routes. In such methods, sections of genes from animals are spliced into the genetic material of bacteria, which then produce the required compound as part of their life-cycle.

Scientists are always looking for ways to chemically modify the structures of pharmaceuticals; one reason is to change the solubility and retention time of the material in the body.

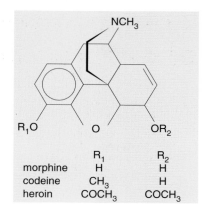

	R_1	R_2
morphine	H	H
codeine	CH_3	H
heroin	$COCH_3$	$COCH_3$

*Fig. 8.3 **The structures of morphine, codeine and heroin.***

Solubility and retention time

Solubility in water and the retention time of a drug in the body are related, though only in a most general way. The more soluble a compound is in water, the faster it is likely to be transported to the liver (the most common site for breakdown or biotransformation of the material), and the sooner it or its more soluble oxidation products can be expelled in the urine via the kidneys. The higher the proportion of hydrophilic groups ($-OH$, $-NH_2$, $-CO_2H$, $-CO_2^-$, etc.), the lower the probable retention time because the material tends to be confined to the blood, lymph and aqueous tissue. The more lipophilic groups there are in the structure (e.g. alkyl side chains $-CH_2CH_2CH_2CH_2CH_2CH_2CH_2CH_3$), the more likely it is to pass into cellular membranes and fatty tissue with slow release into blood or lymph, and the longer is its probable retention time in the body. A balance has to be struck between the two (as well as the other consequences mentioned). If a drug is too lipophilic, it may not be able to achieve sufficient concentration in the bloodstream to be effective or it may not be sufficiently soluble to allow absorption before passing through the gut.

The effects of even minor changes of structure on property are quite marked. Aspirin is sufficiently water-soluble to be an effective pain-killer. The much less water-soluble methyl ester, oil of wintergreen (Figure 8.4), is used for external application (in embrocations and rubs), where lipophilic character is desirable if the material is to be absorbed through the skin.

The solubility of aspirin is increased by preparing the sodium or calcium salt. This, however, is merely to facilitate 'taking the aspirin'; soluble aspirin dissolves in water quite easily. Not only is the solubility greater but also the rate at which it dissolves is increased by mixing calcium carbonate and citric acid in the tablet: this rapidly disperses the tablet in (warm) water. The ionic form $-CO_2^-$ is largely converted back to the unionised $-CO_2H$ in the acidic conditions of the stomach, but by then the aspirin is in a relatively large volume of liquid at 37°C and the decreased solubility is less important. The low pH in the stomach can, however, represent a serious problem. When a drug is seriously affected by acid, it must either be modified for oral use or be given by injection.

The penicillins are important antibiotics that combat certain types of bacteria by preventing them building cell walls. This does not kill the existing bacteria, but when they attempt to divide they are unable to produce viable offspring.

THE COMMERCIAL IMPORTANCE OF ORGANIC COMPOUNDS

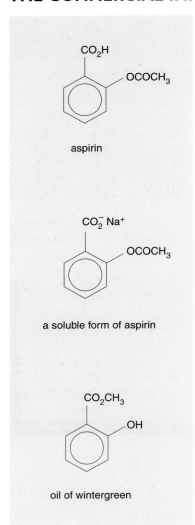

Fig. 8.4 Structures of aspirin, its soluble sodium salt and oil of wintergreen.

Fortunately, mammalian cells are different – they have cell membranes but no walls around them. Since it was first used, half a century ago, penicillin has been modified hundreds of times, and many forms of it are used today.

The original 'penicillin', penicillin G, is extremely water-soluble and rapidly excreted in the urine. During the first clinical trials (in 1941), a policeman's life was one of the first to be saved by its use. It is so soluble, and was in such short supply, that some of the drug had to be recovered from the urine of other patients in order to provide enough.

The solubility in water can be reduced by incorporating lipophilic side chains. Thus the replacement of the sodium ion by procaine gives a species with much greater retention time.

penicillin G

procaine (cation)

In contrast to the need to reduce the hydrophilic character of natural penicillin, it is of first importance that penicillamine (Figure 8.5), a penicillin derivative, is as water-soluble as possible: with two hydrophilic groups and only five carbon atoms per molecule it could hardly be otherwise. This compound's therapeutic use lies in its solubility and the rapidity with which it can be excreted. It is used to form coordination complexes with heavy metals such as copper and lead, e.g. in cases of lead poisoning.

Another example of a drug, in which the hydrophilic or lipophilic character may be increased or decreased by minor chemical modification, is chloramphenicol. The marked hydroxy group may be replaced by hexadecanoate (palmitate):

$$-O_2CCH_2CH_2CH_2CH_2CH_2CH_2CH_2CH_2CH_2CH_2CH_2CH_2CH_2CH_2CH_3$$

to lower the solubility. (It has a nasty taste and can then be administered by mouth as a fine 'insoluble' suspension to patients who cannot swallow a capsule. It is then hydrolysed to chloramphenicol in the stomach and intestines.) Replacement of the same group by the more hydrophilic $-O_2CCH_2CH_2CO_2^-$ Na^+

chloramphenicol

renders it sufficiently soluble for intravenous injection (for patients who cannot swallow at all).

Nitrogenous fertilisers

Nitrogen uptake by plants is essential for the production of proteins and nucleic acids. While the bulk of the plant is non-nitrogenous cellulose and water, enzymes and co-enzymes (ATP, NADH) are necessary for the metabolic processes of the cell, and RNA and DNA are necessary for control and reproduction.

The nitrogen can only be taken up by green plants (unaided by symbiotic bacteria) as inorganic nitrate, NO_3^-, and fertilisers like potassium nitrate (natural), calcium nitrate and ammonium nitrate (both synthetic) are obviously advantageous to plant growth. Indeed, potassium nitrate contains essential K^+, and ammonium nitrate (Figure 8.6) contains nitrogen in both the cation and the anion.

Forms other than nitrates, e.g. ammonium sulphate, are less efficient. The ammonium ion must be oxidised by soil bacteria. Before this can happen it may be washed from the soil by the action of rain. However, the nitrate produced is delivered more slowly. The conversion of organic compounds also involves both hydrolysis and initial reduction to ammonia: this must be even less efficient and slower.

What, then, are the drawbacks of inorganic fertilisers? The two main objections, both of which can be overcome with care, are as follows:

- The nitrogen is delivered in one dose – if too large, it can cause excessive growth, which may encourage foliage rather than flowers and fruit (Figure 8.7).

- Local concentration of the fertiliser can cause burning and shrivelling.

An objection, not related to crop growth, is that the great solubility of nitrates and ammonium salts can more easily result in 'run-off' during heavy rain polluting water-courses. This can lead to excessive plant growth in rivers (Figure 8.8). When the plants die, the bacterial breakdown lowers the oxygen level of the water, with death of higher forms of aquatic life (eutrophication).

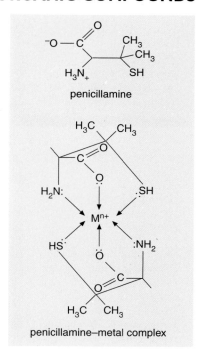

Fig. 8.5 Structures of penicillamine and its metal complex.

Fig. 8.6 Ammonium nitrate pellets and the machinery needed to spread them over the field.

Fig. 8.7 Plant growth is affected by the amount of fertiliser: too little, just right, or too much?

THE COMMERCIAL IMPORTANCE OF ORGANIC COMPOUNDS

QUESTION

What ion present in ammonium sulphate causes its aqueous solution to be acidic? Explain your answer.

QUESTION

During the second half of the twentieth century, Israel developed an extensive agricultural industry. River water (and artesian water), containing minute amounts of naturally dissolved minerals, has been used for irrigation. Recently, in such areas, crop yields have fallen. Suggest why.

Fig. 8.8 A hazard of nitrogenous fertilisers – eutrophication. Algae bloom in a lake polluted by fertiliser, stifling other forms of life.

A particular objection to ammonium nitrate and ammonium sulphate is that, being the salts of a weak base and strong acids, their solutions are acidic.

Plants take up water by osmosis and capillary attraction. The water passes from a low concentration of solutes outside the roots to a higher concentration of solutes in the plant. For those not familiar with osmosis, it sometimes seems more logical to consider the water passing from a high external concentration of water to a lower one inside the plant.

The evolution of plants took place long before the use of artificial fertilisers. The environmental water (essentially rain water) in which they evolved was of a low solute concentration.

The osmotic pressure of an aqueous solution is a measure of its tendency to take up water when separated from water by a membrane through which only this solvent will pass. It is directly proportional to the concentration of particles in solution. Solutes with a high formula mass give rise to solutions with a lower osmotic pressure than solutes with a low formula mass (at the same mass concentration and temperature). Ionic solids, for which one mole of solid gives rise to two (NH_4^+ NO_3^-) or three ((NH_4^+)$_2$ SO_4^{2-}) moles of ions (particles) in solution have, correspondingly, twice or three times the osmotic pressures of solutions of covalent solids of the same formula mass. Plants take up water quite slowly unless deprivation has caused them to droop and increase the concentration of cell fluids: then, when it rains, they may take up water more quickly.

A local concentration of solutes in the environmental water, especially of inorganic salts of low formula mass, leads to a situation where uptake of water cannot occur or, worse, may even be reversed.

Inorganic fertiliser must be delivered with care. Salts adhering to foliage and wetted by dew or irrigation can 'burn'; their leaves are dehydrated and their cell walls may be ruptured by reversed osmosis. Local lumps of such fertilisers can produce too high a concentration, when dissolved, around roots.

With the exception of urea, H_2NCONH_2, organic fertilisers are of biochemical origin. They vary from expensive dried blood or hoof-and-horn products, from abattoirs, to farmyard manure and slurry (Figure 8.9) and domestic compost heaps. Apart from urea, they have a low nitrogen content, the dried blood being the richest. The combination of high molecular mass and low solubility cannot cause any osmotic problems. Indeed, at certain times of the year, farmers put

Fig. 8.9 There may be more nitrogen in the bags than in the 'muck-spreader'.

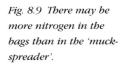

down huge amounts of material on pastures with no significant detriment to the grass (other than temporary top-growth damage by loss of light).

Apart from the low nitrogen content, the disadvantage of such materials is that they must be broken down by bacteria. Against this must be set the improvement in the soil by the breakdown of other organic matter, the encouragement of worm activity, and the advantage of a slow and steady release of nitrogen over an extended period.

An intermediate position is taken by urea. This synthetic organic fertiliser is very soluble in water and contains 47% nitrogen, higher than any other fertiliser applicable by normal means. Its molecular mass, 60, is of the same order as the average molecular masses of the ions in inorganic fertilisers ($\frac{1}{2}K^+NO_3^- = 50.5$; $\frac{1}{2}NH_4^+NO_3^- = 40$), and so osmotic effects are similar. But, unlike ammonium nitrate, its aqueous solution is neutral.

Urea has the advantage over other organic fertilisers that it releases its nitrogen slowly by hydrolysis:

$$H_2NCONH_2 \ + \ H_2O \ \rightarrow \ 2NH_3 \ + \ CO_2$$

Its high solubility makes it susceptible to being washed away by rain.

Esters, oils and fats

Edible oils and fats are naturally occurring esters, mainly of glycerol, propane-1,2,3-triol. Fats usually come from land animals. Oils usually come from marine animals and from the vegetable kingdom. The lower melting point oils tend to replace fats in marine animals, e.g. whales, because fats would be too viscous or solid at sea temperatures.

The carboxylic acids involved, often known as fatty acids, are mostly straight chain acids with an even number of carbon atoms in the range C_{14} to C_{20}; the even number, which includes the carbon atom in COOH, reflects their biogenesis from ethanoate. Milk, butter and dairy products contain a high proportion of esters of shorter acids.

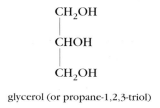

glycerol (or propane-1,2,3-triol)

It is generally true to say that fats (suet, lard, dripping) contain mainly saturated acids such as hexadecanoic (palmitic) and octadecanoic (stearic), while oils (olive oil, maize oil) contain a high proportion of unsaturated acids such as oleic acid. An exception is coconut oil which contains very little unsaturated material.

$C_{17}H_{35}CO_2H$	stearic acid	$CH_3(CH_2)_{16}CO_2H$
$C_{15}H_{31}CO_2H$	palmitic acid	$CH_3(CH_2)_{14}CO_2H$
$C_{17}H_{33}CO_2H$	oleic acid	$cis\text{-}CH_3(CH_2)_7CH{=}CH(CH_2)_7CO_2H$

The glyceride esters (or glycerides) are often known as stearin, palmitin and olein. This adherence to old naming practices is hardly surprising when you consider that olein might properly be called propane-1,2,3-triyl tri(cis-octadeca-9-enoate).

$$CH_2OCO(CH_2)_7CH=CH(CH_2)_7CH_3$$

$$CHOCO(CH_2)_7CH=CH(CH_2)_7CH_3 \qquad \text{'olein'}$$

$$CH_2OCO(CH_2)_7CH=CH(CH_2)_7CH_3$$

propanone   propylbenzene

(CH₃COCH₃)

The higher melting point of the fats is primarily caused by the relative ease with which saturated chains can pack together, maximising intermolecular attraction and melting point. If the unsaturated chains were *trans-* , their packing ability would not be significantly different, but they are largely *cis-*. As you have seen in cyclohexane and cyclohexene, it is customary to represent larger molecules in organic chemistry as carbon skeleton structures in which each end point or junction represents a carbon atom unless labelled otherwise. (This is not suitable for very small molecules; methane would be a dot, ethane a line and ethene an equals sign.)

You can picture the problem of packing when we represent three acids in this way:

stearic acid

a *trans*-octadecanoic acid

oleic acid

The problem of packing is made worse because the natural fats and oils are always mixtures. This is likely to depress the melting point much more for a mixture of esters of unsaturated and saturated acids than those of saturated acids alone.

The use of oils and fats in food is well known; the subject of margarine will be expanded upon. Margarine was invented in the nineteenth century as a cheap substitute for butter with better keeping qualities. The first margarine was obtained from animal fats by a largely mechanical separation of the unsaturated esters which were then 'ripened' with a little milk. (It went sour!). During the twentieth century it has been manufactured by the partial catalytic hydrogenation of oils, e.g. 'groundnut' oil, using a nickel catalyst, all of which must be removed. Removing the unsaturation raises the melting point:

$$\underset{\text{alkenoic ester}}{\overset{H}{\underset{A}{\diagdown}}C=C\overset{H}{\underset{B}{\diagup}}} + H_2 \xrightarrow[\text{Heat and pressure}]{\text{Ni}} \underset{\text{alkanoic ester}}{ACH_2CH_2B}$$

Emulsifiers, yellow colouring, 'butter' odours (an early one was butanedione or 'diacetyl' – you may have some in your laboratory), common salt, antioxidants (which are oxidised by aerobic bacteria sacrificially) are added and the mixture is ripened with skimmed milk. Cheap margarine early in the century led to some dietary deficiency because of lack of natural vitamins. By law, vitamins A and D are also added to the mixture (which by now is collecting a fine selection of E-numbers).

More recently, the importance of 'polyunsaturates' in the human diet has become an issue. These compounds are esters of unsaturated acids with more than one double bond. The two best known are linoleic and linolenic (you do not need to learn their formulae)

Notice that the double bonds are not conjugated and that they are *cis* . Not only would these valuable materials be lost during hydrogenation, but there is concern that at high temperatures the double bonds will become *trans* and may also wander into conjugation. A way of avoiding the problem when manufacturing 'dietary' margarines, said to be beneficial for those with diseased coronary arteries, is to complete the hydrogenation using monounsaturates and then blend in the oils which are richer in the polyunsaturates.

Lubricating oils and essential oils
Lubricating oils are based on the products of vacuum distillation of the higher boiling fractions from petroleum. They are not esters.

Essential oils is a term used rather imprecisely to describe strongly smelling and volatile oils derived from plants. Many are derived from the molecule $C_{10}H_{16}$, limonene, itself an essential oil derived from lemon grass. If they are related to this molecule they are classed as terpenes. They can be aldehydes (citral from orange and lemon peel), alcohols (geraniol and linalol from roses and lavender) or ketones (ionone from violets). If you have some of these in the laboratory you will be very disappointed by their odours; blending is often responsible for the final smell of a flower or a perfume. There are hundreds of terpenes; their ability to isomerise is nothing less than gymnastic! They are the basis of the multi-million dollar perfume industry and most of them are now synthetic. Their study is outside the scope of A-level but you should realise that the volatility and odour of alcohols can be varied by esterification, often with ethanoic acid. Quite simple esters can begin to copy the odour of more complicated ones and they are often used in cheap flavouring, e.g. pentyl ethanoate is often known as 'pear drops'.

THE COMMERCIAL IMPORTANCE OF ORGANIC COMPOUNDS

Soaps

The alkaline hydrolysis of fats and oils gives rise to soaps. The name saponification means soap-making. Usually the alkali is sodium hydroxide though potassium hydroxide is occasionally used in making toilet soaps. Lithium hydroxide is not suitable; the corresponding lithium salts are used in greases for automobiles.

$$
\begin{array}{c}
CH_2OCOR \\
| \\
CHOCOR \\
| \\
CH_2OCOR
\end{array}
\quad + \; 3NaOH \longrightarrow \quad
\begin{array}{c}
CH_2OH \\
| \\
CHOH \\
| \\
CH_2OH
\end{array}
\quad + \quad 3RCO_2^-Na^+
$$

fat or oil   glycerol   soap

The soap is precipitated from the mixture by saturating the solution with common salt. After washing out the excess alkali the crude soap may be mixed with some of the glycerol (produced in the same reaction) to soften it. It is then coloured and perfumed.

The manufacture of soap has declined since World War II, because its use for washing clothes has been replaced by synthetic detergents. In 'hard' water, soaps give insoluble calcium and magnesium salts which form a scum. The corresponding compounds of the detergents (often sodium salts of sulphonic acids) are much more water-soluble.

Polystyrene

This polymer, the recommended systematic name of which is poly(phenylethene), is commonly used in two forms. A hard, high-density form is often seen as the transparent yellow handles of screwdrivers, and a low-density expanded form is familiar as an unbelievably light packing material (Figure 8.10).

The advantage of the latter form stems from its ability to be expanded by gas bubbles to many times its original (true) volume and, at the same time, to be moulded into the shapes of the items it is intended to protect. The packing material hardly increases the weight, and therefore the cost of transport, but it combines crushability and rigidity in such a way as to be damaged sacrificially by impact. It is also packed around objects for transport in the form of pellets.

The inclusion of air or gas bubbles in the structure also makes it an excellent thermal insulator. It is used for ceiling tiles, the interior of refrigerator doors and casings and between partition walls of houses. Unfortunately, it is highly flammable, but its ease of ignition may be partially reduced by the addition of non-flammable matter during the expansion process.

Styrene is often mixed with an inert material as the bulk component of car body fillers. This is mixed with a tiny quantity of 'catalyst', strictly an initiator, often an organic peroxide. The mixture hardens by polymerisation. The polystyrene binds the inert material into a hard solid.

Fig. 8.10 Low-density polystyrene is an excellent thermal insulator.

poly(phenylethene)
(polystyrene)

Condensation polymers

Condensation polymers are those which appear to have been formed by the elimination of water or some other simple substance, such as ammonia or hydrogen chloride, between successive monomers (Figure 8.11).

In the simplest cases the monomers are bifunctional and produce linear molecules. The materials are usually of high tensile strength and make good fibres. Because the reactions by which they are formed are usually reversible, however, unlike the polyalkenes, which have continuous carbon chains, they are often not resistant to chemical attack.

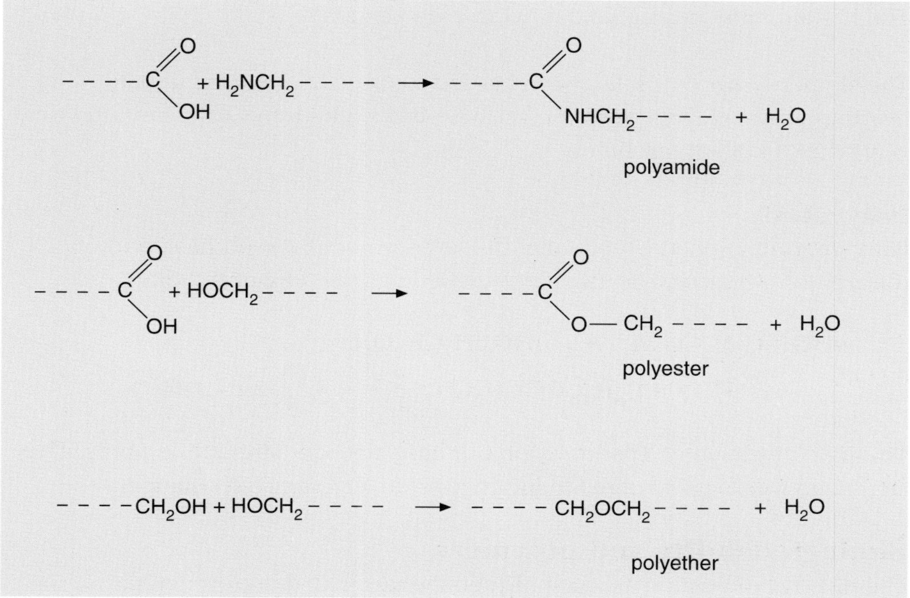

Fig. 8.11 Formation of polyesters, polyamides and polyethers by condensation.

Nylons

These polyamides can be made by polymerisation or co-polymerisation depending on whether one monomer (with both an amino and a carboxy group) or two monomers (one with two amino and one with two carboxy groups) are used. Thus co-polymerisation of 1,6-diaminohexane and hexane-1,6-dioic acid (adipic acid) gives nylon-6,6 (the numbers reflect the numbers of carbon atoms in the amine and acid molecules):

$$n\mathrm{H_2N(CH_2)_6NH_2} + n\mathrm{HO_2C(CH_2)_4CO_2H}$$
$$\rightarrow \mathrm{H[NH(CH_2)_6NHCO(CH_2)_4CO]_{\mathit{n}}OH} + (2n-1)\mathrm{H_2O}$$
nylon-6,6

One of the most widely used nylons is nylon-6 (i.e. one monomer with six carbon atoms per molecule). This is obtained, in theory, by polycondensation of an amino acid with $\mathrm{H_2N-}$ at one end of the molecule and $\mathrm{-CO_2H}$ at the other. In practice, it is obtained from the cyclic amide, caprolactam, by ring opening and polyaddition:

$$\underset{\text{caprolactam}}{\underset{}{\text{H}_2\text{C}\begin{array}{c}\text{CH}_2\text{---}\text{CH}_2\\ \\ \text{CH}_2\text{---}\text{CH}_2\end{array}\begin{array}{c}\text{C}=\text{O}\\ | \\ \text{NH}\end{array}}} \quad \xrightarrow{\text{heat}} \quad \underset{\text{nylon-6}}{-(\text{HN(CH}_2)_5\text{CO})_n-}$$

Nylons are widely used for fibres. They may be used as monofilament (extruded and stretched but unspun) for fishing lines, strimmer cords and toothbrush bristles, or they may be spun into yarn (Figure 8.12), perhaps with threads of a natural fibre . The natural fibres trap air better, and the garments feel warmer: they are also more moisture-absorbent and feel less 'clammy'. The harder, more wear-resistant nylon gives durability.

The slippery nature of solid nylon has led to its extensive use for plain bearings on small carts, lawn-mower wheels and children's toys, etc., and for gear wheels in light machinery, e.g. clocks.

Polyesters

Polyesters are typified by terylene, the co-polymer of terephthalic acid (benzene-1,4-dicarboxylic acid) and ethylene glycol (ethane-1,2-diol):

$$n\text{HO}_2\text{CC}_6\text{H}_4\text{CO}_2\text{H} \quad + \quad n\text{HOCH}_2\text{CH}_2\text{OH}$$
$$\rightarrow \quad \text{HO(OCC}_6\text{H}_4\text{COOCH}_2\text{CH}_2\text{O})_n\text{H} \quad + \quad (2n-1)\text{H}_2\text{O}$$

These polymers are similar to nylon but have superior strength as fibres. They are mixed with natural fibres to improve wear- and crease-resistance

Biodegradability and polymers

The disposal of waste synthetic polymers, particularly those used as packaging material, has caused much environmental concern. Ideally, recycling would make use of what are, in effect, products of the limited resources of the Earth's petroleum. In practice, the time and the workforce required for collecting and sorting plastic materials, and the cost of cleaning and recycling them, are normally very uneconomic when compared with the cost of new polymers.

The problem would be lessened, at least for the present generation, if the polymers could be made biodegradable. Natural polymers such as wool and silk (which are proteins – co-polymers of many amino acids) or wood and cotton and hence paper (polymers of glucose) are sometimes inconveniently biodegradable. Indeed, industry spends millions of pounds each year on chemicals to protect carpets from insect damage and wood from fungal action.

Polyalkenes are among the most difficult polymers to degrade. The hydrocarbon chain is unwettable (bacteria and moulds need water to live and carry out their chemistry) and it offers no chemical point of attack. Hydrocarbon chains *can* be broken down by aerobic bacteria. Butter soon goes rancid because attack occurs near the carboxy group. Margarine and lard are slightly more resistant because of their longer chain length. Soap is slowly biodegradable.

Fig. 8.12 Some nylon products.

Fig. 8.13 Biodegradable plastic cutlery.

THE COMMERCIAL IMPORTANCE OF ORGANIC COMPOUNDS

During the fabrication of polythene bags, carriers, and film used for agricultural 'mulch', the brutal mechanical treatment of the liquid polythene results in the breaking of some of the hydrocarbon chains. This produces short-lived radicals. The extruded hot polythene film comes in contact with oxygen (in air), and peroxide radicals are formed. These break down to give a few keto or carboxylic groups which are potential points for bacterial attack. This may be a way forward in the quest for easily biodegradable plastics but, at present, the necessary increase in temperature and pressure on extrusion makes fabrication more difficult.

Rubber is a natural unsaturated hydrocarbon polymer and 'perishes' very easily if not protected; approximately 20% of its carbon-to-carbon links are double bonds. The present method of ensuring a limited life for polythene sheet in the environment is to incorporate some additional unsaturated material in its manufacture. Exposure to air and sunlight causes oxidation of the double bonds and the resulting carbonyl-containing groups are targets for bacterial attack. The polythene is given a guaranteed minimum life by incorporating some antioxidants (e.g. BHT) which must be 'sacrificed' before oxidation of the incorporated alkene groups can begin.

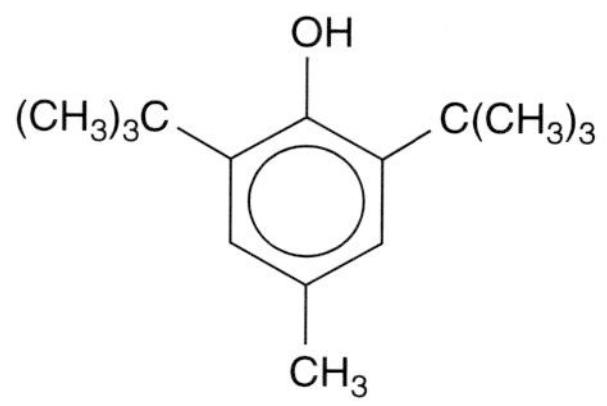

BHT: an oxidant

Assessment questions

Past Edexcel examination questions and part-questions, Unit 5 and synoptic. Students should be aware that these questions cannot include anything on new topics (e.g. NMR, phenols).

1   (*i*)   Complete the boxes below to show the electron configuration of a titanium atom and a Ti^{3+} ion.

		3d					4s
Ti	(Ar)						
Ti^{3+}	(Ar)						

[2]

(*ii*) Titanium is a transition metal. State what is meant by this on the basis of the electron configuration given in (*i*).   [1]

(Edexcel June 1997)

2   Transition elements characteristically have several oxidation states; iron commonly has oxidation states of $+2$ and $+3$, but these are not the only ones.

(*i*)   Explain why, under normal conditions, iron(III) is more stable than iron(II).   [2]

(*ii*) Oxidation of iron with potassium nitrate in alkaline solution gives a red compound, R, which contains potassium, iron and oxygen only. R is a powerful oxidising agent and will oxidise iron(II) to iron(III).

25.0 cm^3 of a 0.100 mol dm^{-3} solution of R oxidised 37.5 cm^3 of a solution of iron(II) containing 55.6 g dm^{-3} of $FeSO_4.7H_2O$.

Calculate the oxidation state of the iron in R.   [5]

(*iii*) Suggest a formula for R.   [1]

(Edexcel Jan 1997)

3   A student determined the percentage of iron in an iron(II) salt, X, as follows:

1.20 g of the salt was placed in a beaker and 50 cm^3 of water was added to dissolve the salt. The solution was heated to boiling point with a small quantity of concentrated nitric acid and eventually it became a yellow colour.

One drop of this yellow solution was then extracted and tested by adding one drop of a freshly prepared dilute solution of potassium hexacyanoferrate(III). Potassium hexacyanoferrate(III) solution will react with iron(II) ions to give a blue coloration; no such coloration was obtained.

Further treatment of the solution included the addition of ammonia solution, a little at a time, with constant stirring, until a slight smell of ammonia persisted. The brown precipitate which had formed was allowed to settle. This precipitate was separated, washed and then heated strongly to give iron(III) oxide.

A further 1.20 g sample of X was treated in the same way, and in both experiments the mass of iron(III) oxide obtained was 0.245 g.

(*a*) Suggest a colour for the iron(II) salt solution.

(*b*) Why was the concentrated nitric acid added?

(*c*) What ion gave rise to the yellow colour in the solution after boiling with concentrated nitric acid?

(*d*) (*i*) The formula of potassium hexacyanoferrate(III) is $K_3Fe(CN)_6$. Write the formula of the complex anion in this compound.

　　(*ii*) Draw a diagram showing the shape of the anion in (*i*). Describe the shape.

　　(*iii*) Why was the hexacyanoferrate(III) test carried out?

　　(*iv*) If a blue coloration had appeared in the test, what would the student then have to do?　　**[6]**

(*e*) Explain why ammonia solution was added until a slight smell of ammonia persisted.　　**[2]**

(*f*) (*i*) Name the brown precipitate formed on addition of the ammonia solution.

　　(*ii*) Write an ionic equation showing the formation of this precipitate　　**[2]**

(*g*) Write the equation for the action of heat on the precipitate.　　**[1]**

(*h*) Calculate the mass of iron present in 0.245 g of the iron(III) oxide.　**[2]**

(*i*) Calculate the percentage of iron present in X.　　**[2]**

(Edexcel June 1996)

4 (*a*) (*i*) Enter the electronic configurations required in the boxes below.

		3d				4s
Cr	(Ar)					
Cr^{2+}	(Ar)					
Cr^{3+}	(Ar)					

　　(*ii*) How is the electron structure of Cr unusual and why?　　**[4]**

(*b*) (*i*) Calculate the oxidation number of Mn in MnO_4^{2-}.

　　(*ii*) Why does Sc form only Sc^{3+} ions?

　　(*iii*) Why are transition metals able to show a variety of oxidation states?　　**[6]**

(c) Fe^{2+} ions will catalyse the reaction between peroxodisulphate ions and iodide ions, which is otherwise slow; the uncatalysed, exothermic reaction is:

$$S_2O_8^{2-} + 2I^- \rightarrow 2SO_4^{2-} + I_2$$

and the catalysed steps are:

$$2Fe^{2+} + S_2O_8^{2-} \rightarrow 2SO_4^{2-} + 2Fe^{3+}$$

$$2Fe^{3+} + 2I^- \rightarrow 2Fe^{2+} + I_2$$

(i) Sketch reaction profiles (enthalpy level diagrams) for both the uncatalysed and the catalysed reactions. Label each profile.

(ii) Examine the equations for the catalysed steps of the reaction. Suggest why this process is more energetically favourable.

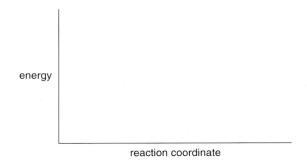

energy

reaction coordinate

(iii) On what property of the transition metal does the catalysis depend? **[6]**

(Edexcel Jan 1996)

5 (a) Give the electronic configurations of the vanadium atom and the V^{2+} ion.

[2]

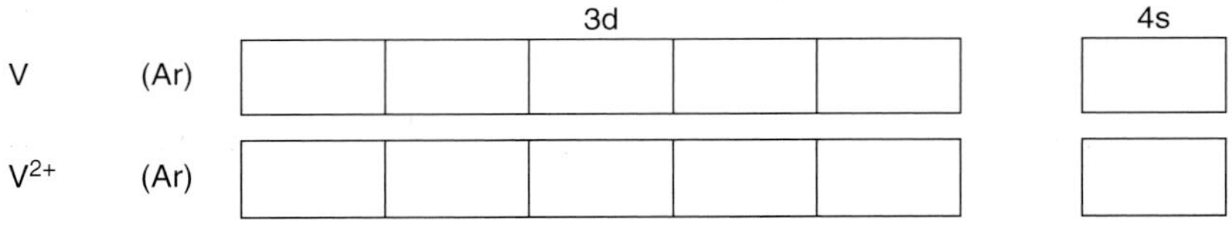

		3d					4s
V	(Ar)						
V^{2+}	(Ar)						

(b) (i) Suggest why the hydrated ion $[V(H_2O)_6]^{2+}$ is coloured.

(ii) Name the types of bonding within ions of this type. **[3]**

(c) Ammonium vanadate, NH_4VO_3, dissolves in aqueous sodium hydroxide with the evolution of a colourless gas. The solution becomes yellow after acidification. The gas has a pungent odour and produces a pale blue precipitate with copper(II) sulphate solution. The precipitate dissolves, as more gas is passed in, to give a deep blue solution.

(i) Write an ionic equation for the reaction of the cation in NH_4VO_3 with alkali.

(ii) Name the pale blue precipitate.

(iii) Give the formula of the ion responsible for the colour of the deep blue solution.

(*iv*) Ammonium vanadate, on treatment with sulphuric acid, gives a yellow colour due to the $[VO_2]^+$ ion. Addition of zinc to the solution causes the solution colour to change to blue, then green, then violet. Give the oxidation number of vanadium in the vanadium-containing ions in each coloured solution.

Blue solution   Green solution   Violet solution   **[5]**

(*d*) The industrial production of sulphur trioxide from sulphur dioxide and oxygen is catalysed by vanadium(V) oxide. It has been proposed that the first stage of the reaction is

$$SO_2 + V_2O_5 \rightarrow SO_3 + 2VO_2$$

Write an equation for the second stage, thus showing vanadium(V) oxide as a catalyst.   **[1]**

(*e*) Give the systematic name for each of the ions:

(*i*) $[VO_2]^+$

(*ii*) $[Cr(NH_3)_4Cl_2]^+$ **[2]**

(*f*) Draw and describe the shape of the ion in (*e*)(*ii*).   **[2]**

(Edexcel Jan 1996)

6   (*a*) Define oxidation in terms of electron transfer.   **[1]**

(*b*) State three properties which distinguish transition metals from main group metals.   **[3]**

(*c*) Give the electron configurations for Fe and Fe^{3+} in the table below:

		3d					4s
Fe	(Ar)						
Fe^{3+}	(Ar)						

(*d*) (*i*) Name the types of reaction involved in the following changes:

$[Cu(H_2O)_6]^{2+}$ $\xrightarrow{\text{aqueous } NH_3}$ blue precipitate $\xrightarrow{\text{aqueous } NH_3}$ deep blue solution
     A   **B**   **C**

(*ii*) Give the formula of compound B.

(*iii*) Draw the structure of the ion responsible for the colour in solution C and show its shape.   **[5]**

(*e*) Addition of aqueous copper(II) ions to aqueous iodide ions gives a precipitate of copper(I) iodide and liberates iodine; the iodine can be titrated with aqueous sodium thiosulphate. These reactions form the basis for a volumetric analysis of copper, for example in metal alloys.

ASSESSMENT QUESTIONS

(*i*) Write an ionic equation for the reaction of copper(II) ions with iodide ions.

(*ii*) Given that the reaction of iodine with sodium thiosulphate is

$$I_2(aq) + 2S_2O_3^{2-}(aq) \rightarrow 2I^-(aq) + S_4O_6^{2-}(aq)$$

find the volume of sodium thiosulphate solution of concentration 1.00 mol dm^{-3} needed to react with the iodine liberated by copper ions from a brass screw of mass 2.00 g, containing 60% of copper by mass. **[4]**

(Edexcel June 1995)

7 (*a*) When an acidified solution of potassium dichromate(VI), $K_2Cr_2O_7$, is added to a solution of an iron(II) compound, the dichromate(VI) ions are reduced to chromium(III) ions and the iron(II) ions are oxidised to iron(III) ions.

(*i*) Write an ionic half-equation for the reduction of dichromate(VI) ions in acidic solution. **[1]**

(*ii*) Write an equation for the reaction between dichromate(VI) and iron(II) ions. **[2]**

(*b*) A 0.204 g sample of steel reacted with dilute sulphuric acid. The resulting solution required 27.4 cm^3 of 0.0220 mol dm^{-3} potassium dichromate(VI) solution for complete reaction. Calculate the percentage of iron in the sample of steel. **[4]**

(Edexcel Jan 1999)

8 This question concerns the compound Q, pent-3-en-2-ol

$$CH_3\ CH\ CH = CH\ CH_3$$
$$|$$
$$OH$$

(*a*) Give the structural formula of a compound isomeric with Q which does not contain a carbon-carbon double bond. **[1]**

(*b*) Q shows two types of stereoisomerism.

(*i*) Draw the geometric isomers of Q, and state why Q shows this type of isomerism. **[2]**

(*ii*) Draw the optical isomers of Q and state what effect such isomers have on the plane of plane-polarised monochromatic light. **[3]**

(*c*) Give the mechanism for the reaction of bromine with Q: you may represent the compound as >C=C< for the purposes of this question. **[3]**

(*d*) (*i*) Give the structural formula of the product when Q reacts with potassium dichromate(VI) solution in the presence of dilute sulphuric acid. **[1]**

(*ii*) Describe a test to show that the product formed in (i) is not an aldehyde. Suggest a positive test which would give the identity of the new functional group that is present in the molecule, and state what you would see. **[4]**

(*e*) The iodine number is a measure of the degree of unsaturation of a molecule. It relates to the number of C=C double bonds in the molecule. It is defined as the number of grams of iodine, I_2, that react with 1.00 g of the compound.

Show that the iodine number of Q is approximately 3 by calculating the number of grams of iodine, I_2, which react with 1.00 g of Q. **[3]**

(Edexcel June 1999)

9 Iodine reacts with propanone in the presence of an acid catalyst according to the equation:

$$CH_3COCH_3 + I_2 \xrightarrow{H^+} CH_3COCH_2I + HI$$

Data concerning an experiment to determine the rate equation for this reaction are given in the following table:

Relative concentrations			
$[CH_3COCH_3]$	$[I_2]$	$[H^+]$	Relative rate
1	1	1	2
1	2	1	2
2	1	1	4
1	1	2	4

(*a*) (*i*) State why the rate equation cannot be written from a knowledge only of the chemical equation representing the reaction. **[2]**

(*ii*) Use the data to deduce the order of reaction with respect to:

propanone   iodine   hydrogen ions **[3]**

(*iii*) What does the order of reaction with respect to iodine tell you about the part that iodine plays in the rate determining step of the reaction? **[1]**

(*iv*) Write the rate equation for the reaction. **[1]**

(*b*) (*i*) Use the bond enthalpies given below, in kJ mol^{-1}, to calculate the enthalpy change for the iodination of propanone:
E(C–H) = 413; E(I–I) = 151; E(C–I) = 228; E(H–I) = 298 **[2]**

(*ii*) Draw an enthalpy level diagram for the reaction, showing on it both catalysed and uncatalysed pathways. **[3]**

(Edexcel June 1999)

10 (*a*) Write an equation to show the reaction that occurs when hydrogen chloride dissolves in water. **[1]**

(*b*) The following electrode (reduction) potentials are required in this question:

$$O_2(g) + 4H^+(aq) + 4e^- \rightleftharpoons 2H_2O(l) \quad E^\ominus = +1.23 \text{ V}$$

$$Cl_2(aq) + 2e^- \rightleftharpoons 2Cl^-(aq) \quad E^\ominus = +1.36 \text{ V}$$

$$F_2(g) + 2e^- \rightleftharpoons 2F^-(aq) \quad E^\ominus = +2.87 \text{ V}$$

(*i*) State and explain what occurs when fluorine is bubbled into water. **[3]**

(*ii*) Write an equation for the expected reaction of chlorine with water using the above data. **[1]**

(*iii*) In fact, the reaction between chlorine and water is usually represented by the equation.

$$Cl_2(g) + H_2O(l) \longrightarrow H^+(aq) + Cl^-(aq) + HOCl(aq)$$

Suggest why this reaction occurs rather than the one you suggested in (*ii*). **[2]**

(*c*) (*i*) Write an equation to show the disproportionation of the chlorate(I) ion in solution and give the conditions necessary for this to occur. **[2]**

(*ii*) Use the equation to explain the word 'disproportionation'. **[2]**

(*d*) When solid $KBrO_3$ is heated, oxygen is evolved, and a white solid remains. If the white solid is dissolved in water and aqueous silver nitrate is added, a cream precipitate is formed which is soluble in concentrated ammonia solution.

(*i*) Suggest the name of the white solid and write an ionic equation for the reaction that occurs on the addition of the aqueous silver nitrate. **[2]**

(*ii*) Hence write an equation to represent the thermal decomposition of $KBrO_3$. **[1]**

(*e*) KIO_3 reacts with hydrogen peroxide according to the equation:

$$2IO_3^-(aq) + 2H^+(aq) + 5H_2O_2(aq) \rightarrow I_2(aq) + 6H_2O(l) + 5O_2(g)$$

(*i*) Deduce the ionic half-equations for the reduction reaction and the oxidation reaction. **[3]**

(*ii*) What would be seen if the reagents were mixed in a conical flask? **[2]**

(Edexcel June 1999)

11 Consider the following reaction scheme:

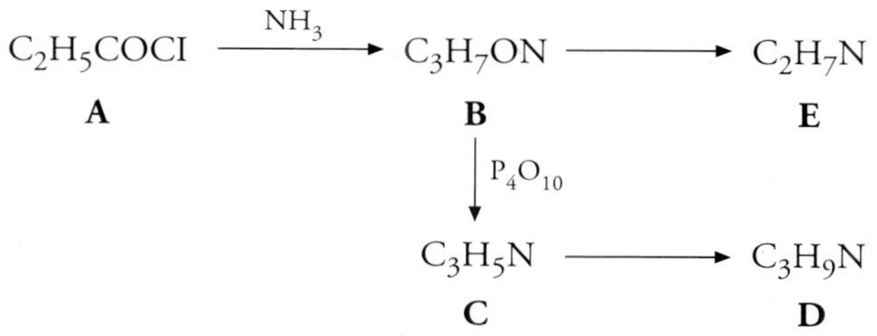

$$C_2H_5COCl \xrightarrow{NH_3} C_3H_7ON \longrightarrow C_2H_7N$$
$$\quad\;\; \textbf{A} \qquad\qquad\quad \textbf{B} \qquad\qquad\;\; \textbf{E}$$

$$\textbf{B} \downarrow P_4O_{10}$$

$$C_3H_5N \longrightarrow C_3H_9N$$
$$\quad\; \textbf{C} \qquad\qquad\; \textbf{D}$$

Compounds **D** and **E** have the same functional group.

(a) Identify, using structural formulae, the compounds **B** and **C**, and the functional group in both **D** and **E**. **[3]**

(b) (i) Give the reagents and conditions necessary for the conversion of compound **C** into compound **D**. **[2]**

   (ii) Give the reagents and conditions necessary for the conversion of compound **B** into compound **E**. **[2]**

(c) (i) Give an equation to represent the reaction that takes place when compound **C** is boiled with dilute hydrochloric acid. **[2]**

   (ii) State the type of reaction taking place in (i). **[1]**

(d) (i) What structural feature of the group present in both **D** and **E** enables them each to react with dilute hydrochloric acid? **[1]**

   (ii) Using the structural formulae, give an equation to represent the reaction of compound **D** with dilute hydrochloric acid. **[2]**

(e) A compound C_6H_7N, having the same functional group as **D** and **E**, reacts with nitrous acid in the presence of hydrochloric acid. The resulting solution reacts with alkaline 2-naphthol to give a red precipitate.

   (i) Give the equation representing the reaction of C_6H_7N with nitrous acid, using structural formulae. **[2]**

   (ii) Show the structure of the product which is formed with 2-naphthol. **[1]**

   (iii) What is the significance of compounds of this type? **[1]**

(Edexcel June 1999)

12 The kinetics of a reaction are used to clarify reaction mechanisms. An experiment to determine the kinetics of the substitution reaction between 2-chloro-2-methylpropane and sodium hydroxide uses equal initial concentrations of these substances in aqueous ethanol solvent. A mixture was maintained at 25°C, and samples were taken at intervals. The samples were quenched in about twice their volume of cold propanone, and the concentration of sodium hydroxide was found.

Time/min	0	7	15	27	44	60
Concentration/mol dm^{-3}	0.080	0.065	0.054	0.041	0.028	0.020

(*i*)   Show, by means of a suitable graph, that the reaction is first order.   [4]

(*ii*) As performed, the results cannot distinguish between the rate laws

rate = k[OH$^-$] and rate = k[halogenoalkane]

Outline a further experiment which must be performed to enable the distinction to be made, showing how the new data would be used to establish the rate law.   [3]

(*iii*) The reaction is, in fact, first order with respect to the halogenoalkane. Write the mechanism for the substitution reaction, identifying the rate-determining step.   [4]

(*iv*) Nucleophilic substitution is usually accompanied by elimination as a competing reaction. Write the name and structural formula of the product of the elimination reaction with 2-chloro-2-methylpropane and state the conditions which favour elimination over substitution.   [3]

(Edexcel June 1999)

13 Consider the following reaction scheme in which all of the compounds have straight chains.

$$A \xrightarrow[\text{STEP 1}]{\text{Mg}} CH_3MgBr \xrightarrow[\text{STEP 2}]{C_4H_8O} \underset{\textbf{B}}{C_5H_{12}O} \xrightarrow[\text{STEP 3}]{K_2Cr_2O_7/H_2SO_4} \underset{\textbf{C}}{C_5H_{10}O} \xrightarrow[\text{STEP 4}]{I_2/NaOH} \underset{\textbf{D}}{\text{Yellow ppt.}} + \textbf{E}$$

(*a*) (*i*)   Give the structural formula of **A**.   [1]

(*ii*) State the conditions under which the reaction in step 1 is carried out.   [2]

(*b*) After giving consideration to the sequence of reactions in steps 2, 3 and 4,

(*i*)   write the structural formula of C_4H_8O;   [1]

(*ii*) write the structural formula of **B**.   [1]

(*c*) (*i*)   Identify the yellow precipitate **D** formed in step 4.   [1]

(*ii*) Give the structural formula of **E**.   [1]

(*d*) Give the structural formula of **C**.   [1]

(*e*) The mass spectrum and infra-red spectrum of **C** are shown below.

(*i*)   Mass spectrum of **C**.

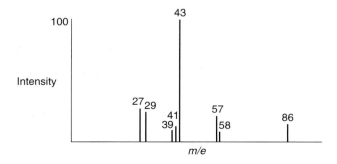

Referring to the mass spectrum, suggest the identities of the species giving rise to the peaks at *m/z* values of 43 (two species) and 86. **[3]**

(*ii*) Infra-red spectrum of **C**.

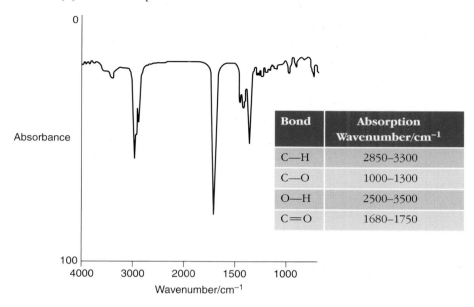

Bond	Absorption Wavenumber/cm⁻¹
C—H	2850–3300
C—O	1000–1300
O—H	2500–3500
C=O	1680–1750

Explain how the infra-red spectrum is consistent with the identity of **C**. **[2]**

(Edexcel Jan 1999)

14 An oily liquid **Z** can be produced from propene by the following scheme:

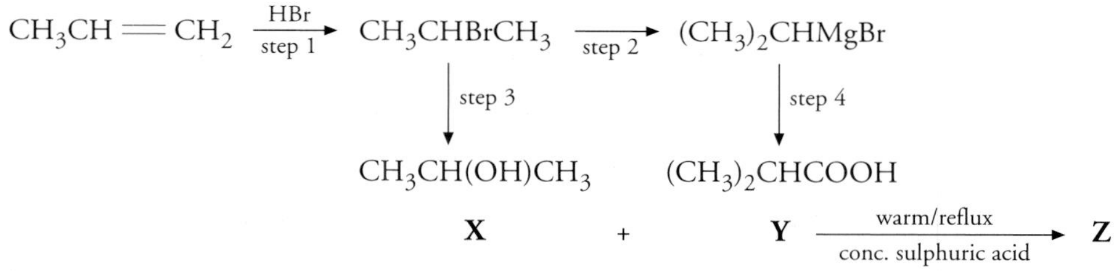

$$CH_3CH = CH_2 \xrightarrow[\text{step 1}]{HBr} CH_3CHBrCH_3 \xrightarrow[\text{step 2}]{} (CH_3)_2CHMgBr$$

step 3 ↓   step 4 ↓

$$CH_3CH(OH)CH_3 \qquad (CH_3)_2CHCOOH$$

$$\mathbf{X} \qquad + \qquad \mathbf{Y} \xrightarrow[\text{conc. sulphuric acid}]{\text{warm/reflux}} \mathbf{Z}$$

(*a*) Write the mechanism for the reaction between propene and hydrogen bromide. **[3]**

(*b*) Give the names of the reagents and the conditions for:

    (*i*) Step 2 **[3]**

    (*ii*) Step 3 **[2]**

(*c*) Step 4 takes place in two stages. Name the reagents for each stage. **[2]**

(*d*) Give the full structural formula of the liquid **Z** formed by the reaction of **X** and **Y**. **[2]**

(Edexcel Jan 2000)

ASSESSMENT QUESTIONS

15 Bulletproof vests are made from Kevlar. The first step in its manufacture is the polymerisation of two monomers:

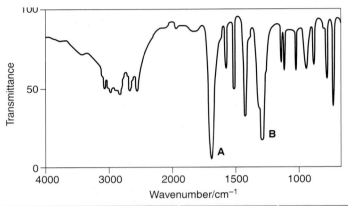

benzene-1,4-diamine   benzene-1,4-dicarboxylic acid

(a) Draw the structural formula of the polymer showing one repeat unit. **[2]**

(b) The infra-red spectrum of one of the two monomers is given below, together with data for some common IR absorption wavenumber ranges.

Bond	Wavenumber/cm⁻¹	Bond	Wavenumber/cm⁻¹
C—O	1300–1100	C=C (arenes)	1600–1450
C=O	1750–1680	C—H (arenes)	3100–3000
O—H	3500–2500	N—H	3500–3300

Use the data to assign the bonds which cause the peaks **A** and **B** and hence identify which of the monomers produces this spectrum. **[3]**

(c) Benzene-1,4-dicarboxylic acid can be made in the laboratory by heating 1,4-dimethylbenzene (boiling temperature 138°C) under reflux for $1\frac{1}{2}$ hours with an alkaline solution of potassium manganate(VII). The potassium manganate(VII) is reduced to a brown precipitate of manganese(IV) oxide. Concentrated hydrochloric acid is carefully added until the brown precipitate disappears. The mixture is cooled and the benzene-1,4-dicarboxylic acid is filtered off and recrystallised from boiling water.

(i) Why is it necessary to have a reflux condenser in this preparation?

(ii) Why must the solution be cooled before filtering?

(iii) The purity of the product was tested by observing its melting temperature. What would be noticed if the sample were still impure? **[3]**

(d) (i) Poly(1,1-difluoroethene) or PDFE is a piezoelectric substance. This means that when it is physically deformed, it produces an electric current. Draw the structure of the polymer showing two repeat units.

    (*ii*) Name the two different types of polymerisation involved in the manufacture of Kevlar and PDFE. **[4]**

(*e*) Propane and octane are both used as fuels in car engines. 1.00 g of propane (boiling temperature –42°C) produces 2.99 g of carbon dioxide and 50.4 kJ of energy when burnt. 1.00 g of octane (boiling temperature +126°C) produces 3.09 g of carbon dioxide and 48.4 kJ of energy when burnt.

    (*i*) Give one environmental disadvantage of using octane rather than propane as a fuel, stating clearly why its use is more harmful.

    (*ii*) Suggest one disadvantage of using gaseous rather than liquid fuel in a car engine. **[3]**

(Edexcel Jan 2000)

16 Butenedioic acid has the structure

HOOC–CH = CH–COOH

It has two geometric isomers, *cis*- and *trans*-butenedioic acid.

(*a*) Explain in terms of orbitals why butenedioic acid has geometric isomers. **[2]**

(*b*) The following concerns *cis*-butenedioic acid.

    (*i*) Draw the structure of the compound formed when *cis*-butenedioic acid reacts with bromine dissolved in an organic solvent.

    (*ii*) On heating to 145°C, *cis*-butenedioic acid loses water to form a compound containing a five-membered ring which includes one oxygen atom.

        $C_4H_4O_4 \rightarrow C_4H_2O_3 + H_2O$

        Draw the structural formula of $C_4H_2O_3$. **[3]**

(*c*) (*i*) Draw the structure of *trans*-butenedioic acid.

    (*ii*) Suggest why *trans*-butenedioic acid does not lose water at 145°C **[2]**

(*d*) If *cis*-butenedioic acid is reduced with lithium tetrahydridoaluminate(III) (lithium aluminium hydride), $LiAlH_4$, the product is *cis*-but-2-en-1,4-diol:

$HOCH_2$–CH=CH–CH_2OH

    (*i*) State the conditions for the reaction between $LiAlH_4$ and *cis*-butenedioic acid.

    (*ii*) How would you test for the presence of the hydroxy group, and what would you see as a result of your test?

    (*iii*) If *cis*-but-2-en-1,4-diol reacts with the compound $ClCOCH_2COCl$, a polymer results. Draw enough of the chain of this polymer to make its repeating structure clear.

    (*iv*) Suggest why the polymer would not be suitable for use in strongly alkaline or acidic conditions. **[8]**

(Edexcel Jan 2000)

The Periodic Table of Elements

Period

Group

Key

Atomic number
Symbol
Name
Molar mass in g mol^{-1}

Group 1	2											3	4	5	6	7	0
																	2 He Helium 4

Period 1: 1 H Hydrogen 1

Period 2: 3 Li Lithium 7 · 4 Be Beryllium 9 · 5 B Boron 11 · 6 C Carbon 12 · 7 N Nitrogen 14 · 8 O Oxygen 16 · 9 F Fluorine 19 · 10 Ne Neon 20

Period 3: 11 Na Sodium 23 · 12 Mg Magnesium 24 · 13 Al Aluminium 27 · 14 Si Silicon 28 · 15 P Phosphorus 31 · 16 S Sulphur 32 · 17 Cl Chlorine 35.5 · 18 Ar Argon 40

Period 4: 19 K Potassium 39 · 20 Ca Calcium 40 · 21 Sc Scandium 45 · 22 Ti Titanium 48 · 23 V Vanadium 51 · 24 Cr Chromium 52 · 25 Mn Manganese 55 · 26 Fe Iron 56 · 27 Co Cobalt 59 · 28 Ni Nickel 59 · 29 Cu Copper 63.5 · 30 Zn Zinc 65.4 · 31 Ga Gallium 70.4 · 32 Ge Germanium 73 · 33 As Arsenic 75 · 34 Se Selenium 79 · 35 Br Bromine 80 · 36 Kr Krypton 84

Period 5: 37 Rb Rubidium 85 · 38 Sr Strontium 88 · 39 Y Yttrium 89 · 40 Zr Zirconium 91 · 41 Nb Niobium 93 · 42 Mo Molybdenum 96 · 43 Tc Technetium (99) · 44 Ru Ruthenium 101 · 45 Rh Rhodium 103 · 46 Pd Palladium 106 · 47 Ag Silver 108 · 48 Cd Cadmium 112 · 49 In Indium 115 · 50 Sn Tin 119 · 51 Sb Antimony 122 · 52 Te Tellurium 128 · 53 I Iodine 127 · 54 Xe Xenon 131

Period 6: 55 Cs Caesium 133 · 56 Ba Barium 137 · 57 La▲ Lanthanum 139 · 72 Hf Hafnium 178 · 73 Ta Tantalum 181 · 74 W Tungsten 184 · 75 Re Rhenium 186 · 76 Os Osmium 190 · 77 Ir Iridium 192 · 78 Pt Platinum 195 · 79 Au Gold 197 · 80 Hg Mercury 201 · 81 Tl Thallium 204 · 82 Pb Lead 207 · 83 Bi Bismuth 209 · 84 Po Polonium (210) · 85 At Astatine (210) · 86 Rn Radon (222)

Period 7: 87 Fr Francium (223) · 88 Ra Radium (226) · 89 Ac▲▲ Actinium (227) · 104 Rf Rutherfordium (261) · 105 Db Dubnium (262) · 106 Sg Seaborgium (263) · 107 Bh Bohrium (264) · 108 Hs Hassium (269) · 109 Mt Meitnerium (268) · 110 Uun Ununnilium (269) · 111 Uuu Unununium (272) · 112 Uub Ununbium (277)

▲ **Lanthanide elements**

58 Ce Cerium 140	59 Pr Praseodymium 141	60 Nd Neodymium 144	61 Pm Promethium (147)	62 Sm Samarium 150	63 Eu Europium 152	64 Gd Gadolinium 157	65 Tb Terbium 159	66 Dy Dysprosium 163	67 Ho Holmium 165	68 Er Erbium 167	69 Tm Thulium 169	70 Yb Ytterbium 173	71 Lu Lutetium 175

▲▲ **Actinide elements**

90 Th Thorium 232	91 Pa Protactinium (231)	92 U Uranium 238	93 Np Neptunium (237)	94 Pu Plutonium (242)	95 Am Americium (243)	96 Cm Curium (247)	97 Bk Berkelium (245)	98 Cf Californium (251)	99 Es Einsteinium (254)	100 Fm Fermium (253)	101 Md Mendelevium (256)	102 No Nobelium (254)	103 Lr Lawrencium (257)

146

Index

Page references in *italics* refer to a table or an illustration.

INDEX